World Treasury of
Birds in Color

World Treasury of Birds in Color

Engravings by Audubon, Gould and several artists

Texts by Cecil Madsen

Translated by David Macrae

GALAHAD BOOKS · NEW YORK CITY

Men have always felt a great affection for birds, and have always been fascinated by their flashing shapes, their melodious voices, the quickness of their movements and the way they behave.

Birds have more corpuscles in their blood than man because they absorb great quantities of air not only in the lungs, as is the case in mammals, but even in the finest blood vessels of both body and limbs. After all, it is not flight that is the truly distinctive feature of birds, since a number of quadrupeds, such as bats, and even fish, such as the flying-fish, are capable of flight; rather, it is the way they breathe. Birds lack that movable partition known as the diaphragm, which prevents the air from going beyond the chest in mammals; in this way, air reaches into every part of their body, along the respiratory tracts which extend throughout all cell tissue, into the feathers, bones and even between the muscles. In this way, since the air they breathe in dilates and very much lightens their body, they are, in a sense, inflated, and virtually able to swim in air.

Birds use their wings rather like oars, in order to steer and go up or down.

Not all birds can fly; take the ostrich, for example, which has rudimentary wings which it uses to brush the air aside as it walks.

Birds' wings are either pointed or blunt. On pointed wings, the flight feathers gradually diminish in length from the wing-tip to the body: the more pointed the wing, the better the bird will be suited for long flight or sudden, agile maneuvers. The twelve flight feathers which make up the tail act as a rudder, for changes of direction.

Birds combine all forms of motion: they can fly, walk or swim, as their various habits require. Each part of their body, though similar in all species, is specially adapted to their mode of life.

It is interesting to note that, wherever a bird's skin is covered with feathers, the skin itself is very thin; yet in those parts not covered by feathers, it is stronger and may even be covered with scales.

Feathers are made of hornified material consisting of a shaft (at first hollow and then solid), barbs and barbules, these latter being equipped with hooks which join up the barbs. The feathers of the wings and tail are called flight feathers or quills.

The feathers of birds from hot climates are the most brilliant; indeed, the hotter the climate, the more dazzling and colorful are the feathers. In some species, it is the males who wear a splendid display of colors, while the females are clad in dull, somber tones. Yet it often happens that the plumage is the same for both sexes. The young look like the adults only after their first molt.

All birds shed their old feathers at least once a year, in order to put on brighter ones. The molt, as it is called, usually happens in autumn, and occasionally in both spring and autumn. Throughout the molting period, the bird is silent and gloomy. But once this critical period is over, he proudly displays the bright colors which may easily outshine the flowers around him.

Talons vary according to the bird's habits. Those of a bird of prey are hooked and powerful, whereas the claws of walking bird are straight, big and flat. Usually, the claw on the big toe is the strongest, but this is not always so. A claw sometimes found on the wrist of the wing of certain birds is known as a spur: in a numer of species, it can be a dreaded weapon. It may also occur on the tarsus of the foot. The bill consists of two bony sections, called mandibles, which are, in turn, surrounded by a horny substance. Depending on the animal's habits; it may take an endless variety of shapes, being used in some cases to tear into prey, and in others to grind seeds or break a hard object. The bill is the bird's main weapon for attacking a victim or defence against enemies.

Furthermore, it is used to prepare the soft bed in which the young will be hatched.

Together with the tongue, the bill is a mere accessory of the digestive apparatus: the bill is used for seizing and grinding, and the sole purpose of the tongue is swallowing.

The digestion of birds is so active that some of them can get fat in an extraordinarily short time. For example, thrushes and garden buntings can become quite plump after only five or six days.

Food first reaches a swelling in the oesophagus, called the crop, or first stomach, where it remains while undergoing certain changes to facilitate digestion. When it passes into a second stomach, the *proventriculus*, the food is treated with gastric juices. It finally turns into chyme in the third stomach, or gizzard, which is equipped with powerful muscles; it is able to handle the most solid objects and can grind even small stones.

Curiously enough, it is possible for seeds entering the digestive tract to pass through it quite intact and then be dropped in areas where circumstances will allow them to grow. The presence of unusual species of trees in a given region may thus be accounted for in this way.

The sense of touch, smell, taste and hearing are not highly developed in birds. It is true that some historians have been impressed by the fine sense of smell of birds of prey, who are capable of finding their way to a corpse on a battlefield in a matter of hours. But more scientifically-minded travellers like Audubon and Levaillant have done experiments which prove that such birds are guided mainly by their vision.

Indeed, sight is highly perfected in birds, more so, in fact, than in any other class of animals. First, the eye occupies a relatively large volume within the head. It contains a special organ, apparently found only in birds, the *pecten*, which is generally square and lamellated. It is a black, folded membrane, well equipped with blood vessels, lying at the back of the eye-socket, but inclined towards the cristalline lens. Anatomical science has so far failed to explain its use, but it is thought to enable the bird to see far and near by advancing or withdrawing the cristalline lens. The other parts of the eye, such as the choroid, the iris, the retina have no remarkable features. The white of the eye is surrounded by a circle of bone or cartilage which forms a hard protective ring.

Besides the two ordinary eyelids, one upper and the other lower, birds have a third one, which is really an extended fold of the conjunctiva. This transparent fold, situated vertically, covers the eye like a curtain, protecting it from excessively bright light. It is this eyelid, or nictitating membrane, which enables the eagle to look straight at the sun, and nocturnal birds of prey to avoid becoming dazzled when they have to face bright daylight.

The perfection of birds' vision is dramatically demonstrated when one sees a vulture, appearing as a mere speck in the sky, suddenly swoop down on to the prey it has detected from enormous heights; or when one sees a swallow able to distinguish and catch tiny insects while flying at great speed. According to Spallanzani, the swift has such penetrating vision that it can spot an object of only five lines in diameter at more than three hundred feet.

Birds are the only creatures in the world capable of covering vast distances astonishingly fast. Whereas the best runners among the mammals can barely manage 13–15 miles an hour, certain birds can easily travel fifty miles in the same time. Big birds, such as the kite or the eagle, measuring more than a meter long, can take off and be quite lost to sight in less than three minutes. These birds must be assumed to cover about a mile a minute.

The vocal apparatus of birds is quite complicated and different from that of man. It consists of a kind of bony chamber, which is really a swelling of the trachea at the point where it branches off, on entering the chest, to form the bronchial tubes. This lower larynx is the singing organ of birds.

Birdsong is an expression of their feelings. Birds sing both for their own pleasure and also in order to delight anyone who may happen to be listening. While they are filling the woods with melodious sounds, they seem to take a conscious delight in having their voices admired, rather like an artist on stage: they look around in order to attract attention.

Birds vary their song according to the season; but their graceful tones and harmonious concerts are especially admirable in the early days of spring. Can there be anything more pleasant than the chirping of the warbler among the bushes at dawn; or the measured phrases of the nightingale, poetically breaking the stillness of the woods, on a calm June night?

Birds certainly have a language which is understood by them alone. At the approach of danger, just one bird lets out a particular cry which serves as a warning to all the other individuals of the same species, who then hide until their fears are dispelled. The plaintive whistle of a blackbird, which means that a bird of prey is coming, is enough to make any bird within earshot stand perfectly still.

Reproduction takes place in birds at regular times fixed by nature. Birds are remarkable for their faithfulness in love. It is quite common for a male to become attached to a female and live with her until one of them dies.

In spring the birds, working in pairs, set about gathering the materials necessary for the building of a nest, bringing in perhaps a twig or a piece of moss. Big birds are usually satisfied with a crudely built nest: a few bits of wood or branches interwoven in the middle of a bush. But the smaller species display extraordinary craftsmanship in shaping a charming miniature basket which they then line with wool, horsehair or down. The male and the female work together, as they want their eggs to rest on a soft, warm and solid bed. The mother thinks up countless ingenious ways of hiding the nest from unwelcome eyes: she may put it in a bush or on a forked branch, at the foot of a tree or in the hollow of a tree, on chimney-stacks, roofs and against walls, etc. Nests of the same species are always made in the same way.

When the nest is finished and when the bird has cemented its walls with a kind of putty, made of clay and a viscous saliva secreted by glands under his tongue, the eggs are laid.

The smaller the bird in a given species, the more eggs are laid. For example, the eagle lays only two eggs, whereas the tit lays from 15 to 18.

Most species of birds make periodic journeys. These migrations are almost always so regular that they have given rise to a sort of natural calendar. They are either annual and regular, or irregular and accidental, that is to say caused by necessity or as a result of some atmospheric disturbance.

It is quite common to see numerous flocks of birds, guided by a leader, cover huge distances, cross seas, move from one continent to another at remarkable speeds. At the right moment and when the wind is favorable, they set off, in the Old World, towards the south-west in autumn and the north-east in spring; in America, towards the south-east. These wandering flocks always know to find their way back to the right area; sometimes each female lays its eggs in the same nest.

How can such small creatures travel such enormous distances, stopping only at certain places for food, and managing without sleep throughout the whole long and tiring journey? How, for example, can the quail venture across the Mediterranean

twice a year? It remains a mystery.

　We would still owe birds a great debt of gratitude, even if all they did for us was to brighten our lives, delight our ears and flatter our eyes.

　Yet they in fact provide us with exquisitely tasty food, which is held in high esteem, as well as a soft, delicate down. More-over, birds are useful to man because they destroy the insects, grubs and caterpillars which are a constant danger to crops. Without them, agriculture would be impossible.

　In the whole of Creation they are an order which is at once indispensable and wonderful.

1 – Perching Birds

Chaffinch

Chaffinches live in flocks, but tend to scatter when looking for food. They occur throughout Europe, in North Africa and in various regions of Asia, where they are either residents or transients. Their diet consists of a variety of seeds; their habitat may be in woods, gardens or near high mountain peaks.

The chaffinch's song can be heard at the beginning of spring. They are remarkably gay little birds; in fact, proverbially so, at least in French.
(A)

Buntings

Buntings may be divided into two broad categories, depending on whether the claw of their big toe is short and hooked or long and straight.

The first category includes the Reed Bunting, which is highly typical of Buntings as a whole, the Cirl Bunting and the Ortolan Bunting, well known to hunters and gourmets. The second category includes the Snow Bunting, which lives in the northern areas of Europe.
(B)

Bullfinch

The Bullfinch is a pretty bird, with a red breast, which wears a grey mantle and a black skullcap. It feeds on seeds, berries and buds. People like to keep it as a pet in a cage, because of its happy disposition, its devotion to its master and the ease with which it learns to sing. When living in the wild state, it builds a very nicely designed nest in the form of a bowl.   (C)

(B)

(B)

(C)

(A)

Linnets

Linnets are very sociable birds, staying together all the year round, sometimes in very large flocks, except at breeding time. Eggs are laid about two or three times a year. The males take no part in the building of the nest or the brooding; but they do look after their females and bring them food.

Linnets feed mainly on hemp seed and linseed, whence their name; at the end of winter they also pick at the buds on the trees. Their song is sweet, brilliant and varied. Since they are also docile and affectionate, people like to keep them as cage-birds. Several species are to be found, all over Europe and North America. (A)

Siskin

The Siskin's overall length is only 5 inches. Its range is almost the whole of Europe, but it is a migrant, reaching the southern regions in autumn flying in large flocks. It flies at high altitudes. Though not very melodious, its song is sweet and varied. The siskin is readily tamed, and willingly eats its favourite delicacy, hempseed, out of a man's hand. It does well in captivity, as it is not a quarrelsome bird; its lively nature and its year-round singing seem to have a stimulating effect on any other birds that may be with it in an aviary. (B)

Goldfinch

The elegant Goldfinch is one of the prettiest species in Europe. It has a brown back, with a red face and a beautiful yellow patch on the wing. It has a pleasant voice and is very docile. When raised in a cage or an aviary, it becomes very friendly, learns to sing and even to do certain exercises. Goldfinches can sometimes be seen pulling along little bowls containing their food and drink, pretending that they are dead, and playing other games. (C)

(A)

(B)

(C)

Greenfinches

The distinctive features of the Greenfinch are: a bill almost as long as its head; round nostrils, hidden under its frontal feathers; thick tarsi; long toes and very sharp, hooked talons.

The Common Greenfinch lives in forests. Its large nest is built of moss and dried grass on the outside, and is lined with horse-hair, wool and feathers on the inside. It sets up a kind of food stockpile around its nest.

The female incubates her 5 or 6 eggs with great care. The male is also a devoted family bird, and can often be seen hovering around the tree in which the nest is situated, accompanying his acrobatics with a gay chirping. On his way to and from the nest, he emits two curious sounds, which, in German at least, have given him a name meaning "bell". Greenfinches are easily tamed and quite mild, and can even learn a few words. They can be taught to eat off one's finger, answer their master's voice, etc. (A)

Hawfinch

This bird has an astonishingly stout and strong bill, not at all commensurate with its small body-size, which is about the same as that of a blackbird. It feeds on seeds, berries and, sometimes, on insects; even the hardest kernels cannot withstand the force of the powerful tool with which it is equipped. Hawfinches are to be found throughout Europe; according to the temperature, they live in open country or in the woods. They are nasty and quarrelsome; for this reason they cannot be kept in a cage with other birds, as they will attack and kill them.

In the Americas there are several species of crested hawfinches, which have a beautiful red plumage. (B)

Cardinal

The *Cardinal* gets its name from the bright scarlet coloring which is characteristic of most of its species. They have this color over most of the body, with a black face, and have a mobile red crest on the top of their head. Young and female cardinals have more somber tones. Their voice is loud and even penetrating though they also make a pleasant chirping noise throughout much of the year. Cardinals can be raised in cages, feeding on seeds. They are residents of North America. (C)

Crossbill

This bird's mandibles are curved in opposite directions, and meet about two thirds of the way along its rather short, stubby bill. Its sturdy legs are armed with powerful, very slightly curved talons which are almost triangular in shape. As one might expect, a bird of such an unusual design has a rather unusual life-style. It lives mainly in pine forests and plantations of resinous trees, feeding on the seeds which it extracts from their cones. It is most common in northern Europe and in North America. (D)

(C)

(A)

(B)

(C)

(D)

Tits

The perching-birds belonging to the tit family have a short, straight and tiny bill, with little hairs at its base. Their really distinctive traits are to be found in their habits, such as their liveliness, audacity, courage, and remarkably sociable way of life. The tit is the bird which discovers owls during the day, harasses them noisily, pecks at them and forces them to run away, pursued by a swarm of rebellious smaller birds, all mobilized by its example. The tit is a tough little fighter: it gives free rein to its instincts whenever it gets a chance, making up for its lack of strength by the daring of its assaults. This bird is movement personified. It is continually flitting to and fro among the branches of the trees, looking for food, poking its bill under the bark, or hanging from the branches with its feet in the air, trying to catch the insects feeding on the inner surface of the leaves.

Its diet changes according to the seasons and the circumstances in which it finds itself. In addition to a very wide range of insects, including bees and wasps, it also eats seeds and fruits with both soft and hard outer coverings. It can even be carnivorous, as it often kills weak or sick birds and eats their brains. Certain species are fond of tallow and rancid fat.

Tits have a very strong attachment to each other; so much so that a flock of them would sooner be decimated or even completely wiped out rather than abandon a wounded comrade. In spring, however, they withdraw in pairs, to breed.

The location of the nest varies according to the species. Some build them deep inside cracked walls, in hollow trees or abandoned nests. Others nest in trees, or hang their nest from the ends of branches; yet others hide theirs among the reeds. Their building techniques are usually highly ingenious.

Tits abound throughout Europe, in Asia and in North America. The most noteworthy species are the *Great Tit*, the *Blue Tit*, the *Crested Tit*, the *Long-Tailed Tit*, the *Marsh Tit* and the *Penduline Tit*.

House sparrows

The *House Sparrow* is found throughout Europe from far North to deep South.

Every knows this lively, bold and cunning little bird, which lives in flocks in residential areas, and even in the middle of the big cities. They are familiar birds, but their familiarity is tempered with caution; they always take care to stay a safe distance from man.

Sparrows look for food in small groups, and nest fairly near each other either in cracked walls, under the eaves, in trees or even swallows' nests which they have taken over for themselves. Two or three times a year they lay between four and eight eggs, on a bed of feathers and soft grass. As can be seen, their fertility is very high. They can eat a wide range of foods, but prefer seeds and insect grubs.

Much has been written to show that sparrows do terrible damage to crops and should, accordingly, be exterminated; it has also been argued, at equal length, that they should be preserved, on account of their great appetite for caterpillars. Now it is generally felt that they are useful. (A)

Widow-Bird

The habits of the *Widow-Birds* are in no way exceptional. These small perching birds were given their name on account of the dark patches all over their plumage. The tailfeathers of the male are extraordinarily long. (B)

(A)

(B)

Warblers

The *Reed Warbler* weaves its nest among reeds, which act as pillars and supports. The *Fan-tailed Warbler* builds a nest in the form of a purse or distaff made of wool, spider's webs or other silky materials suspended in the midst of a cluster of reeds or other marsh grasses. The *Redheaded Tailorbird* is the most astonishing of all. It uses its bill and feet to make cotton taken from a special tree into strands; it then chooses the strongest and broadest leaves, makes holes in them, and stitches them together with the fibers it has prepared, thus making a kind porch which hides the nest perfectly from its enemies. Unlike the species we have just described, this bird is not found in Europe: it lives in India and neighboring islands.

The other commonest species are: the Garden Warbler and the Blackcap. Also the Subalpine Warbler, which lives in the high Alpine plains.

The *Reed Warbler*, the *Great Reed Warbler*, *Blyth's Reed Warbler* and the *Marsh Warbler* live in marshlands, the wooded or grassy areas near ponds or rivers, and damp gardens. They can be seen constantly climbing up and down the branches of trees and shrubs with great agility. One usually sees these birds at funny angles, and hardly ever standing on something level. They are quarrelsome birds, who resent the presence of other animals. They are thought to have much in common with the nightingale, as the male sings both day and night, while the female is brooding. The various Red Warblers and the Marsh Warbler may not be able to match the beauty of the nightingale's song, but their own song certainly carries quite a long way; they often accompany their song with a fluttering of their whole body. These birds mate in a very delicate way, usually on the frail leaf of a reed. Their nest is quite artistically made, and well padded on the bottom. It is deep enough to prevent the 4 or 5 eggs from rolling out when the reeds bend in the wind. Nests of this sort have been almost touching the surface of the water in a strong wind, yet the birds inside did not fall out. (A/B)

Goldcrests and Firecrests

These are the smallest European birds. Their diet consists of the smallest insects, but they prefer the tiniest earthworms most of all, sometimes gorging themselves on them. They are to be found most commonly in oaks, elms and tall pine trees. They are constantly in motion, restlessly flitting from one branch to another. These birds can be tamed quite quickly, and soon take food from a man's hand. Their nest, usually located at the end of a branch, is round and is made of delicate moss, of grub cocoons and thistledown. Goldcrests and Firecrests are found in many countries outside Europe. (C)

(B)

(A)

(A)

(A)

(C)

Tree-Creepers

The familiar *Tree-Creeper* has a total length of 5—6 ins. It is common in most of Europe, Siberia and North America. This lively little bird is found in woods, orchards, trees with dense foliage and near streams. It spends most of its time climbing around the trunks of trees. It has a thin, high-pitched cry, which is the only way one can tell it is there, as it constantly hides on the side of the tree furthest from the person pursuing it. It is accordingly very hard to observe. It nests in holes in trees, lining the nest with grass and moss bound together with spider webs. (A)

Nuthatch

The *Nuthatch* is about the size of a sparrow, with grey-blue upper parts, rust below, and a white throat. It is resident throughout most of Europe, living in pairs, usually in oak forests. It spends a part of the winter hidden in a tree hollow, the entrance to which it partly seals off with wet earth. Beginning in February, it devotes its entire day to the search for insects on nearby branches; it drags them out of the bark, using its strong, pointed bill which can make quite a large hole in a matter of minutes. It continually emits a high-pitched, monotonous call. (B)

Wall-Creeper

Whereas many birds spend much of their lives in the trees, this one climbs, nests and lays its eggs on walls or sheer rock-faces. According to one writer, it is particularly fond of cemeteries, and has been known to lay its eggs inside human skulls. It flies with a beating motion of its wings. Flies, ants and spiders make up its diet. Though the wall-creeper is so tame that it will not fly away even when someone comes within a few feet of it, it nevertheless will stop what it has been doing, when inspected at close range, and stare back at the visitor. (C)

(B)

Ravens and Crows

These birds, even more so than related species, will eat anything: live prey, carrion, stranded fish, insects, eggs, fruit, seeds. Their depredations are enormous. For example, the raven, not content with taking its toll of the moles, fieldmice and young hares of the countryside, invades the farmyard and unceremoniously picks off chickens, ducklings and young pheasants. According to Buffon, in some countries, the raven will cling to the back of oxen and devour them piecemeal, having picked their eyes out first. As for the crow, there is no doubt, according to Lewis, that it attacks lambs grazing in the fields in Scotland and Ireland. And crows and ravens alike love to poke around in newly sown earth to feed on the seeds just planted there by the farmer. Not surprisingly, country-dwellers are their implacable enemies, always glad to hunt or trap them.

Out of fairness, however, it must be recognized that ravens and crows do help clean up the environment by devouring carrion which might otherwise infect it; each year, moreover, they destroy vast quantities of worms, grubs and insects.

The flesh of the raven and the crow has a bad smell, caused by the birds' fondness for rotting meat; it is therefore hardly edible. But the rook and the jackdaw, having a different diet, make very suitable game-birds.

Ravens have a strong, sustained flight; their sense of smell and their vision are both highly developed. These two very sharp senses help them, as they soar effortlessly at high altitude, to spot the victims which they contribute to Nature's daily deathroll. Their cry, a rather disagreeable croaking noise, has done much, together with their funereal plumage, to give them a reputation as a bird of ill-omen. The ancients believed that they had the power to foretell the future and, particularly, to predict catastrophes. The Haruspices used to consult them for this reason, and based their utterances on their different ways of croaking and moving through the air.

Ravens and crows are to be found all over the world. The Raven and the Carrion Crow are sedentary, staying in one place all their lives. But the Hooded Crow, the Rook and the Jackdaw are migrating birds which visit the northern countries only at the approach of winter. The Pied Crow is to be found exclusively in Africa. (A/B/C)

(B)

(C)

26

(A)

(C)

Magpies

The features which distinguish *Magpies* from ravens are their shorter wings, their longer, tiered tail, and their less funereal plumage; apart from these slight differences, they are quite similar. Like the ravens, their diet is extremely varied, but does not include carrion. A curious trait of theirs is their compulsive habit of storing things away and picking up any shining objects they see. They have such a powerful instinct for theft that they long been famous for it.

Magpies are distrustful, particularly of man. Yet they will readily attack dogs, foxes and all birds of prey, pursuing them as soon as they appear and usually driving them off, either on their own or through the combined efforts of other magpies responding to their call. These restless birds can run quite briskly, but do not fly well. The continual and sometimes deafening chattering noise they make has become proverbial. The magpie's nest, built at the top of the tallest trees or bushes, is remarkable for its size, design and solidity. It is made of brambles, sticks and sand. This bird starts building several nests at the same time, some quite conspicuously, but the last one with all manner of precautions, so as to avoid being seen.

Magpies lay seven or eight eggs, the male and the female taking turns with incubation. Both have strong parental feelings, and continue to care for their young for a long time.

Magpies can be found all over the world. The European *Magpie (pica pica)*, a common resident of open country throughout Europe, has a beautiful velvety-black color, with its wings and part of its breast a pure white. One species occurring in Brazil and Paraguay has entirely sky-blue plumage, except for a black head and throat. (A)

Jay

The Jays can be distinguished from ravens by the following characteristics: a short bill, slightly indented at the tip, and their ability to make the feathers of their head stand up on end when they are annoyed. They feed mainly on beans, acorns and nuts, but also eat worms and insects, and, like ravens and magpies, they take the eggs and the young of other birds. Although Jays are, by nature, quarrelsome and easily provoked, they are quite easy to tame, and can be taught to pronounce several words. They are common in Europe, America and the Indies. The European Jay is a very pretty species, with small blue feathers at the beginning of its wings. (B)

(B)

(A)

(A)

(B)

Blackbirds and Thrushes

The characteristic feature of the Blackbird and Thrush family is a compressed, curved and slightly serrated bill. It is one of most abundant families of its order, comprising no less than 150 species, profusely represented all over the world.

Birds in this family are usually migratory, and travel in large flocks. They feed on berries, fruits and insects, and have a remarkable gift of song. The division between the two groups is based on their coloring: birds belonging to the Blackbird section of the family have uniform coloring, whereas those belonging to the Thrush section have speckled breasts.

The main species of the first section are: the *Blackbird*, the *Rock Thrush*, the *Blue Rock Thrush* and the *Mocking Bird*.

The *Blackbird's* plumage is a beautiful black color throughout. It lives mostly among brushwood or in coppices near the water's edge. When it finds somewhere with an abundant food supply, it stays there all its life. The Blackbird is highly distrustful and devious; it is rare indeed for man to take one by surprise, unless it has been tempted by its greed into dangers which it otherwise would have avoided. Though a wild bird, it often wanders into public or private gardens, near human dwellings. If taken young, it can grow accustomed to captivity quite easily.

It nests close to the ground, on bushes or in trees. The female works alone at nest-building, while the male entertains her with his song. Four to six eggs are laid, two or three times a year.

The *Rock Thrush* is smaller than the Blackbird, and, unlike it, prefers mountainous regions.

The *Blue Rock Thrush* is easily recognizable by its dark blue plumage. Its habitat is more or less the same as that of the Rock Thrush, as are its habits; but it is much wilder, and has an even more delightful song.

Of all the species in this family, the *Mockingbird* has the most advanced vocal powers; this bird is a native of North America, particularly of Louisiana. Audubon held such a high opinion of the mockingbird's melodious tones that he did not hesitate to rank it far above the nightingale. It, moreover, has the curious ability to imitate, with added embellishments of its own, the songs of all other birds, and even the call of mammals. For this reason, the Indians called it the "Bird with four hundred tongues." (A)

Thrushes

The main species of thrush are: the Common Thrush, the Redwing, the Mistle Thrush and the Fieldtare. Throughout history, the Common Thrush has been prized as a tasty game-bird. The Romans valued this bird so highly that they used to raise thousands of them in huge aviaries, in which poor light was judiciously combined with a suitable diet. Nowadays, no-one takes the trouble to raise thrushes in captivity, since they fatten themselves quite nicely while in the South of Europe in autumn, eating grapes, figs and olives in huge quantities. Their bloated bodies then present an easy target for the shotgun, which claims vast numbers of them. It used to be thought that they got drunk, feeding in the vineyards, whence the French proverb "saoul comme une grive", drunk as a thrush. But in fact that is not the reason why they are unable to flee from danger in autumn; it is solely because they are overweight.

The Redwing has all the qualities and defects of the Common Thrush, and is, like it, highly sought after by gourmets. The two other species are less important from the culinary point of view. All of these species inhabit Europe. (B)

(B)

(A)

(A)

(A)

(A)

Robin

Robins spend the whole of the summer in the woods and approach human dwellings only as they are about to leave in autumn and as they arrive in spring. They build their nest near the ground, on the roots of young trees or on grass strong enough to support it; it is made of moss, woven with bits of horsehair and oakleaves, with a bed of feathers inside; quite often, once they have finished the building of the nest, they heap up a mass of leaves on top of it, leaving only one narrow, inconspicuous opening, which they cover with a leaf when going away. During the breeding season, the woods resound with the subtle, melodious love-song of the male. Robins are not afraid of man and will accompany anyone walking through the woods. (A)

Nightingales

The Nightingale is renowned throughout the world for its song, which is really in a class by itself among the birds of Europe. This wild bird loves to retreat into cool, sheltered places, far from intruding human eyes. Undergrowth, coppices, and thick bushes near the water's edge are its favorite abode. This is where it builds it nest, close to the ground, if not actually on it, and in a rather casual manner. A curious trait of the nightingale is that they sing in the dark, as well as during the day. But the song stops abruptly as soon as anyone comes too close. In fact, nightingales value solitude above all else.

They arrive in Europe alone and also leave alone, in August, for Africa and Asia. (B)

Pipit

This bird is a resident of the South American tropical zone. Its long, very pointed and slightly curved bill is smooth at the sides; it has medium-size wings and a forked tail. It favorite habitat is the tops of the highest trees. It is a gay, lively little bird, which is quite friendly with little birds of other species. (C)

(C)

(A)

(B)

(B)

Swallows and Martins

Swallows and Martins are easily recognizable by their long, pointed wings, their forked tail and their extraordinarily short tarsi. The air is quite literally their element: they fly with astonishing ease, grace and speed. Their whole life is one long flight: they eat, drink and even bathe on the wing. They even feed their young in the air, as soon as they have begun to try out their wings. Their speed is such that some species do about 80 miles an hour.

However, the flight of these birds has developed to such perfection only at the expense of another faculty: walking. Their legs are so short that they find it virtually impossible to walk; indeed, if they ever happen to land, they succeed in taking off again only after the most strenuous efforts. On the other hand, their visual powers are outstanding.

Swallows and Martins are famous for their migrations. Early in spring, they fly in from the warmer climates not in flocks, but individually, or in pairs, and quickly set about repairing their old nests or building new ones if their old ones have been destroyed in the meantime. Among the returning birds, there are always young birds from the previous years who have never nested in Europe. It does seem amazing that these birds, after a six-month absence, are able to find their way back home without the slightest uncertainty. Yet they have been seen doing it too often for there to be any possible doubt on the subject.

The shape, type and location of the nest varies according to the species. The *Swallow* builds its nest against the inner walls of chimney-stacks; the *House Martin*, in the corners of windows and under the eaves of houses. Others nest in dead trees, sometimes in large numbers. Audubon reckoned that there were 11,000 swallows or martins nesting in a huge sycamore which he once observed near Louisville. Some species build their nests in cleft rocks or under the roofs of caves. The *Sand Martin* digs a tunnel, two or three feet deep in steep river banks, and withdraws to the farthest end of it, resting on a bed of feathers.

Most frequently, the nest is made of wet earth mixed with straw, and is lined with feathers and down.

Throughout history, swallows have been privileged to enjoy the sympathy and good-will of men. The ancients regarded these birds as sacred, and even today people still feel a very special tenderness towards them. They are useful birds, in that they destroy vast numbers of insects; they are very gentle and affectionate to each other and, particularly, to their young; they are a good omen, as their coming announces the return of spring: for all of these reasons, they are greatly cherished by Man. (A)

Larks

The characteristic feature of the Lark is its foot. The claw of the big toe is long, strong and straight, sometimes longer than the toe itself, which suggests a walking bird, incapable of gripping the branches of trees and, accordingly, not a perching bird. The lark does, in fact, live on the ground, amongst the harvests covering the great plains, and is very useful to the farmer because of the vast quantities of worms, caterpillars and grasshoppers it destroys every day. It nests in furrows, between two clumps of earth, without much intricacy, but is smart enough to hide its nest from its enemies. It lays 4 or 5 eggs, as often as three times a year when the season is favorable. When the young hatch, after 15 days of incubation, they have another 15 days ahead of them before they are able to leave their cradle. Nonetheless, the mother continues to look after them, to guide their steps and provide for their needs, hovering over them; only when the needs of a new family have to be met does she leave them to their own resources. They will have developed by that time to the point where maternal protection is no longer required, in any case.

After the nesting season, the larks gather together in large flocks, and, with feeding their only concern now, they quickly put on weight; for most of them this is tantamount to a death sentence. Hunters soon appear from all sides to raid these tasty tidbits.

The main species are the Skylark, the Crested Lark, which has a small erectile crest on top of its head, and the Wood Lark, which inhabits the woods during the mating season and perches on the thick branches of trees. (B)

Wagtails

Wagtails habitually follow herds of cattle and flocks of sheep while they are out grazing, without showing any sign of fear. These graceful little birds are the inseparable friend and companion of the man working in the fields. They are quite intimate, even cheeky, with people they know; yet the sight of a stranger is enough to make them fly away. In spite of their great friendliness towards man, they hate to be cooped up in a cage, and never survive long if put in one. However, if they are left free to fly about in an apartment during for the winter, they enjoy it and readily feed on bread crumbs given to them. They have been known to fly on board ships in port, and stay with the sailors throughout the voyage, only leaving them at the other end. (C)

(B)

(A)

(C)

Starlings

Starlings have a straight bill, turned slightly downwards at the tip. They are graceful, lively birds. They have a distinctive dark plumage, with green or blue irridescent tones. They can be seen swarming over the woods or the open countryside in huge flocks. When flying to their resting-place for the night, they maintain a regular formation, whether it be a triangle, rectangle, elliptical circle or sphere. They feed on berry seeds, insects, worms and small land molluscs. For their nests they seek out the most sheltered places, such as hollow trees, cracks in walls, clocktowers, the eaves of houses, and they even go inside dovecotes.

Starlings exist all over the world, and are residents only in certain regions. They are much sought after, not on account of their meat, which has an unpleasant taste, but because of their natural charm and the ease with which they learn how to talk.   (A)

Starlings: Rose-Colored Starling

These birds are similar to Starlings in both form and habits, but their bill is different, being slightly indented. They are very gregarious when feeding and when resting for the night: whole swarms of them can be seen packed onto one tree, sometimes overflowing on to nearby trees as well. These lively, pleasant birds are often seen around residential areas; besides being friendly, they are also useful to man, because they habitually land on the backs of cattle grazing in the fields and eat the bugs that are bothering them. Another useful habit of theirs is their liking for grasshoppers, which they destroy in vast numbers, sometimes even in the form of eggs and grubs.

However, they charge a high price for their help, by eating lots of cherries and mulberries. When the insects on their diet are in short supply, they even eat all kinds of seeds and some cereals.

Rose-colored starlings quickly get used to captivity, just like starlings. They share with them the ability to imitate various words or cries, which is why they are kept as cage-birds in certain parts of India.

The European Rose-Colored Starling has a beautiful pink color on its belly and its back. Other similar species are to be found in Africa, Asia and Java, though these also occasionally show up in Southern Europe on their migrations.   (B)

Waxwings

The Waxwings are sociable birds, living in large flocks throughout the year, except at breeding-time. They feed on buds, berries and insects; they even hunt down flies in flight. They are remarkably lazy birds, and move only when the need for food makes it strictly necessary for them to do so. Most of the time, they remain withdrawn in thick bushes; they are hardly ever seen on the ground, where they would be at a clear disadvantage on account of their clumsy, slow gait. They cannot really be said to have a song; it is more like a faint chirping, which lasts a long time in some species. Waxwings can be found in Europe, North America and Japan. The European species, or *Bohemian Waxwing*, nests in far-northern areas, emigrating to Germany as winter approaches. It is a pretty bird, with a trailing crest on top of its head.   (C)

(A)

(A)

(C)

(B)

(A)

Orioles

The Orioles' bill is long, convex, tough and is equipped with a prominent ridge; the tarsi are very short. They may be found throughout all the warm areas of the Old World, and in the islands of Oceania. The colors of their plumage are very rich, consisting of a variety of subtle combinations of yellow and black.

The *Golden Oriole*, with its beautiful yellow plumage, arrives in May, and leaves again about the middle of August; however, though it arrives alone, it leaves with its family. It nests in big trees, such as oaks and poplars, near the water's edge or the border of the forest. This bird is common in the South of France.

It lays between four and six eggs. It feeds on insects, grubs and caterpillars, and has a particular weakness for various fruits, such as blackberries, cherries and figs. As a consequence of this rather special diet, its meat is considered a delicacy, and it accordingly is a target in the sights of many hunters. The oriole does not get used to captivity, and dies after a few months in a cage.   (A)

Shrikes

The *Shrike* family comprises a number of species with a conical or compressed bill, more or less hooked at the tip and deeply indented; they share with the birds of prey a warlike disposition and a taste for live flesh.

The destructive instinct of these birds is quite astounding. They delight in the bloodshed and death they leave behind them wherever they go. Not content with killing to satisfy their hunger, they seem to enjoy slaughtering insects, birds and small mammals, which they then neatly impale on the thorns of bushes and hedges, as a reserve supply of food.

In spite of their small size, they are not afraid of ravens, magpies and even certain birds of prey, and willingly engage them in battle. They fight bravely to defend their territory against intruders, even against their own species.

Shrikes usually live in big forests, either on the outer edge of the forest, or in secluded places, in the bigger trees or among brushwood. In daytime, they perch on the highest branches, a good vantage point as they look for prey. They are not good fliers, though they glide quite well. They keep up a continuous chatter which is not wholly disagreeable. A striking trait of theirs is their gift for imitation. Indeed, they can imitate any bird-call, and are thought to use this ability in order to lure small birds towards them, thus increasing the number of their victims.

Shrikes emigrate every year. As their body is covered with a thick layer of fat at that time of the year, they become very appetizing targets for sportsmen down below. They are to be found all over the world.   (B)

Flycatchers

The bill of the birds of this family points downwards, is hooked, and has a prominent ridge, with stiff hairs at its base.

Flycatchers feed on winged insects, which they pursue in the air with the most astonishing speed and agility; thier diet sometimes includes caterpillars and ants—indeed it is to catch such insects, and for that reason alone, that they ever come down to the ground. They are silent birds; they fly alone, deep in the forest or at the water's edge, among the reeds and the rushes. They do not sing, even at mating time, and build their nests quite carelessly, without bothering to conceal them from the eyes of their enemies. The location of the nest varies according to the different species, and may be in trees, bushes, cleft walls, wells or under the roofs of houses. The female lays between 3 and 6 eggs, once a year in Europe and as many as three times a year in the other parts of the world.

Flycatchers are migratory birds, and exist in a great number of species all over the world. Europe has several, including the Spotted Flycatcher and the Bec-Figue. This latter species, which is very fond of fruit, is hunted in the South of France for its very delicate meat.   (C)

(A)

(B)

(C)

Drongoes

Drongoes are about the same size as the blackbird, with a shape similar to that of the raven. Their bill is streamlined and very curved; they have a forked tail. Their plumage is black, with irridescent shades of green or blue. They live in small groups in the forests of India, Oceania and Southern Africa, where they inflict great losses on the bee populations. Every morning and evening, they take up their positions, on some dead or leafeless tree at the forest's edge, waiting for the bees to appear, on the way to or from their abodes. Then they pounce from their observation post and slaughter large numbers of these insects.

On account of their rowdy, destructive character, they have been given the name of Bird-of-the-Devil. Drongo meat is not worth talking about; but some species are said to sing quite nicely. The two outermost quills of the Greater Racket-tailed Drongo are long and stringy, ending in the shape of a paddle. (A)

Tanagers

The various species of Tanager all come from the hotter regions of South America. They are remarkable for their conical bill, which is triangular at its base, and also for their brilliant colors. They have roughly the same habits as sparrows or warblers. They are lively, spirited birds who spend most of their time aloft, in trees or bushes, where they feed on berries, insects and seeds; they seldom come down to the ground. Some species live alone while others live in families or flocks. Some of them have a pleasant chirping song. Among the most colorful are: the *Seven-Colored Tanager*, the *Red-Hooded Tanager*, the *Vermilion Tanager*, and the *Masked Crimson Tanager*. The *Palm Tanager*, a remarkably sociable bird, gets its name from its habit of building a huge structure on the tops of palm-trees, to accommodate the nests and eggs of its companions, in a number of compartments. (B)

(B)

(A)

Troupials

The Troupial's bill is as long as its head, thick at its base, stout, straight, with the upper mandible ending in a point on the forehead. These birds, known to the Spaniards as *tordos*, are common in the New World, mainly in Virginia, Carolina, Louisiana and Mexico. They are gregarious and have a pleasant song. Different species of troupials often gather together, sometimes joining birds of other species. Their diet consists mainly of insects and seeds. Although they fly in the open, and can be seen on the ground or perched on lianas or branches of trees, they take great care to hide their nest from view.

2 – Sea Birds

Albatross

The Albatross is the biggest and bulkiest of all the birds which fly over the surface of the sea. They belong to the southern hemisphere. Their wings, when outspread, have been known to reach a span of 16 ft. Their plumage is generally white, only the mantle being black.

But size is not synonymous with courage; this is certainly true of these birds who, in spite of their surprising strength and their large, powerful bill which is both hooked and sharp, are in fact quite cowardly. The much less powerful gulls will attack them and harass them, trying to slash open their belly. The best response the albatross can make to such attacks is to dive under water. Although they themselves are highly voracious eaters, the mere sight of smaller birds such as gulls will make them drop their catch and flee.

They feed on small marine animals, molluscs, mucilaginous zoophytes, and fish eggs and fry. They can swallow quite large fish whole. When they are full, since they have no room for the whole of the fish they have caught, they have to keep part of the fish outside their bill, until the part they have eaten is digested. Several snakes are known to do the same thing. Having gorged itself in this way, the albatross has only one choice open to it if it should be pursued: it has to unload surplus food from its stomach.

These birds have remarkable powers of flight, and travel very far from land, especially during storms. They seem to enjoy being in the midst of the unrestrained elements. They can go for several weeks without sleeping; when they need to rest they simply float on the surface of the sea, with their head hidden under one wing. One can go up quite close to them; in this way, all that sailors have to do in order to catch one is to hit it with a gaff or spear it with a hook. (A)

Petrels

The slightly bulbous bill of the petrels has a hooked part at its tip which looks as if it was hinged on to the rest of the upper mandible. These birds do not dive and rarely swim; but, as they fly, they skim swiftly over the waves, giving the impression that they are running on the water. It is this trait of theirs which led to their being called *Petrel*, or *Little Peter*, an allusion to the miracle performed by St. Peter, when he walked on the rough waters of Lake Genazereth.

The Petrel family contains a number of species, of widely differing size. They have a powerful, rapid flight with which they can cover huge distances, almost always gliding. They come close to the shore only to nest. They lay a single large white egg in a crack in some steep rock; while brooding they make a faint, continuous whirring sound, such as might be made by a small wheel rotating.

On the whole, petrels are not very attractive to look at; yet, for the poor inhabitants of the South Sea islands they are a valuable resource, partly on account of their meat, but mainly because of their warm down and the oil which can be extracted from their stomach.

The amount of oil that these birds carry in their bodies is so great that they can even use it to feed their young. In the Faeroe Islands, this oily substance has been used to make candles for many years.

Petrels love stormy weather: they run across the turbulent waves, and seem to be really enjoying themselves as they follow the mountains of foam. When it gets too rough for them, they fly to safety on a nearby reef or on a ship's rigging. Seeing them appear only during storms, the superstitious, unthinking sailors of years ago used to regard them as evil spirits, the messengers of bad weather. Their blackish plumage only strengthened this feeling.

Petrels find it hard to walk on the ground. If they need to rest while out at sea, they land on in ice-floe, and, with their head under a wing, they allow the wind to take them where it will. (B)

(B)

(B)

(A)

Frigate Bird

The main features of the Frigate Birds are: a bill longer than their head, both mandibles curved at the tip, the front of the neck lacking feathers, very long wings, a long forked tail, feet with incomplete webs.

Frigate birds have a wingspan of 9'9" with a correspondingly powerful flight. They inhabit the intertropical seas in both Eastern and Western hemispheres and sailors claim to have seen them as far as 500—800 miles out to sea. When a storm breaks, they simply climb high above it, and wait in those Empyrean zones for the winds to die down. Their huge wings enable them to stay aloft, without rest, for days at a time.

The vision of these birds is so acute that they can spot columns of flying fish at distances which would render them invisible to a human eye. They drop out of the sky onto their hapless prey, which, for a brief moment, has left its natural element; with neck and feet horizontal, they swoop low over the waves, and seize their utterly astonished victim. Boobies returning from a fishing trip often find their catch grabbed from them by a marauding frigate bird, thus becoming its natural, though involuntary supplier.

Frigate birds have such a warlike disposition and such complete confidence in their own strength that man inspires no fear in them. They have even been seen to fly up to sailors and try to snatch a fish out of their hand.

When they have had their fill of fish and the other marine animals that make up their diet, they fly ashore and perch in a tree, to digest their food in peace.

They gather together in large flocks on the islands which are their customary nesting-grounds. In May, they start repairing their nests or building new ones. They use their bill to cut off small dry branches and make their nest out of them, laying them in a criss-cross pattern. In these nests, which hang from trees leaning over the water, they lay 2 or 3 eggs.

These birds are common in Brazil, Ascension Island, Timor, the Mariana Islands and the Moluccas. They are to be found in tropical zones. (A)

Cormorant

Cormorants have a large, awkward looking body, their feet are short and withdrawn into the abdomen, their head is small and flat and their guttural pouch is very small. The different species vary in size from that of a goose to that of a teal. The blackness of their plumage suggested a link between them and ravens, whence the name Cormorant, or marine crow.

These birds, which may be found all over the world, are always near the sea shore and the mouths of rivers. They are excellent swimmers and divers and fish at very high speed. Their prey hardly ever escapes.

Cormorants always swallow their prey head-first. When they happen to have got hold of the wrong end, they flip it in the air so that it lands head-first in their mouth. It may take a half-hour for it to get an eel into its stomach the right way. The bird can then be seen struggling violently to swallow its prey; just when it seems that the slippery morsel has been downed successfully, it may pop up from its living tomb and fight for its life. The cormorant swallows it again, the eel resists, with its tail sticking out of the bird's mouth. But, eventually, the victim tires of his hopeless struggle and resigns himself to his fate.

The cormorant, a web-footed bird, has an endless appetite; it gorges itself until it can literally take no more. It causes much havoc in the rivers, being able to eat, in a single day, from four to five kilograms of fish.

Since these birds are expert fishers and are easily tamed, in some parts of East Asia they are used domestically. The Chinese and the Japanese have used the fishing ability of these birds more than anybody. After putting a ring around their neck, so that they will not swallow the fish themselves, they release them in areas well stocked with fish. The cormorants, being trained to obey the human voice, bring back their prey.

They have a fast, sustained flight; yet their agility in the water is matched by their clumsiness on land. They are by nature gentle and trusting, and allow themselves to be approached quite readily when resting on trees or among the rocks.

The Cormorant may be found in both the Old and New Worlds: it is a migrating bird. (B)

Gannets and Boobies

The Boobies are large, clumsy-looking birds, bigger than ducks, with white plumage. Rightly or wrongly, they are thought to be stupid, whence the name Booby. A man, encountering one of them in his path, will find that they put up no resistance whatever, and allow themselves to be beaten senseless rather than simply running away. The daring frigate bird makes them disgorge any fish they may have captured. The structure of their bodies accounts for this lack of resistance. Their wings are too long and their legs too short for them to be able to fly quickly out of the reach of their enemies.

Once they are airborne, however, they soar most admirably, with their neck outstretched, their tail spread and their wings almost motionless. Although they can fly quite fast, they rarely go far from land, and are never found more than 50 miles out to sea. In this way, their presence tells a sailor that he is getting close to land. They fly low over the waves, skimming the surface, and grabbing fish, for example herrings and sardines, that are swimming within their reach. The size of their prey is no impediment, as the skin of their throat stretches to accommodate whatever they swallow. In any case, the boobies are quite good divers, capable of staying underwater more than a minute chasing after fish.

Boobies can be found all over the world. They prefer tropical areas, but exist in large numbers in the Hebrides, Norway,

(C)

(A)

(B)

(C)

Scotland and even Kamchatka. When the cold weather approaches, they move south, and, in very severe winters, they pause in Holland and Britain.

There are three known species: first, the Gannet, its Latin name being *Sula Bassana*, from the Isle of Bassan, near the Firth of Forth, on which they gather in large numbers; this is the only European species; then, the Common Booby, or *Sula Dactylatra*, smaller than the gannet, and to be found on Ascension Island; and lastly the Brown Gannet or *Sula Fusca*, from South America.

Pelicans

Pelicans have a long, wide and flat bill. The lower mandible is equipped with a bare membrane which can dilate to become a voluminous pouch. These water-birds live either by the sea or on the shores of rivers, lakes or marshes. When fish reveal their presence, either by the reflections of their scales in the sunlight, or by leaping, the pelican quickly moves in to seize this easy prey. They simply open up their wide bill and scoop up whatever happens to be at hand.

Their appetite—like their stomach——is so vast that they can take, on one single fishing trip, as much fish as would be needed to feed six men. The Egyptians dubbed it the "river camel", as it can absorb about 20 pints of water at one sitting. It eats only two meals a day, but they are no ordinary meals!

Pelicans often fly in groups, near the mouths of rivers or the sea-shore. Having found a suitable place, they begin to beat the surface of the water with outstretched wings, so as to drive the fish ahead of them; as they approach the river bank or a small inlet, they force the fish into a confined space. Their communal meal then begins. When they have had their fill, they go ashore in order to digest their meal undisturbed. Some of them recline, with their neck curled up on their back; others busily preen their plumage. All of them wait patiently for their digestion to do its work and hunger to draw them to yet another banquet. One of them will occasionally empty his well-stocked pouch and gaze admiringly at his catch. Then he swallows it.

This pouch, which is so crucial for the life of the pelican, is made up of two skins, the outer one being an extension of the skin of the neck; the inner one lies next to the wall of the esophagus.

Notwithstanding its great size, the pelican has an easy, sustained flight.

They nest in cleft rocks near the water. Sometimes they simply lay their eggs—usually between 2 and 4 in number—in a hole which they have crudely lined with blades of grass.

After forty or forty-five days of incubation, the young hatch, covered with a grey down. The female feeds them by pressing her pouch against her breast in order to disgorge the fish straight into the waiting bills of the young. This habit is probably at the origin of the fable about the female pelican piercing her own breast, and feeding her young with her maternal blood.

Young pelicans can be quite easily tamed. It is thought that they can be trained, and that they can be taught to fish, like the cormorant.

Pelicans are much more common in warm climates than in the temperate zone. They abound in Africa, Thailand, China, Madagascar, the Sunda Isles, the Philippines, and in the Americas from the Antilles down to the southernmost tip of the continent.

The best-known species are the following: the *White Pelican*, which is about the size of the swan; the *Dalmatian Pelican*, which has white plumage, but black shafts on the feathers of its wings and back; the feathers on its head and the upper part of its neck lie across each other, thus making quite a sizeable crest. It lives in the vicinity of the Black Sea and the islands at the mouth of the Danube; it has also been found in Senegal. Its height is about that of the White Pelican; the *Brown Pelican*, smaller than the first two species, has white and grey about its head and neck; its entire plumage is a grey-brown, with whitish streaks on its back; its pouch is a greyish-blue with reddish streaks. It is found in the Greater Antilles, on the coasts of Peru, in Bengal and in South Carolina; the *Australian Pelican*, found only in the Southern Hemisphere, looks rather as if it is wearing spectacles, because of the ring of bare skin which surrounds its eyes. Its plumage is white. (A)

(A)

Jaegers and Skuas

Jaegers and Skuas have a distinctive very strong bill, which is almost cylindrical and covered with a membrane from its base to the nostrils; the upper mandible is convex, hooked and armed at the tip with a small claw which seems to have been an afterthought.

These birds spend most of their time close to the edge of the sea; but they do go inland after a storm. They fly very fast, even against very strong headwinds. They chase gulls and terns, sometimes even boobies and cormorants quite ruthlessly, in order to seize their prey from them. But it is the gulls and terns which are their usual suppliers. They constantly pursue, harass and attack them, until they disgorge their prey. Even before the fish reaches the sea, they dive after it and catch it. This distinctive habit gave rise to the notion that they fed on gull droppings, whence their Latin title *Stercorarius*.

Jaegers and Skuas almost invariably live alone, so as to assure a regular food supply. They lay between two and four eggs, the male and female each taking turns at brooding. They are brave birds, and will defend their offspring against any animal, even the Man.

There are four recognized European species: the *Parasitic Jaeger*, for which the British name is *Arctic Skua*; this bird lives in Greenland, Newfoundland and Spitzberg, and often visits the ocean coasts; the *Skua*, known in Britain as the *Great Skua*, which breeds in scattered colonies on moors near the sea, in Iceland, Faeroes, Shetlands and Orkneys; the *Pomarine Jaeger*, or *Pomarine Skua*, which is very common in Newfoundland, Ireland and the Faeroes; and the *Long-Tailed Jaeger*, or *Long-Tailed Skua*. (A)

Gulls

These birds abound at the sea's edge, where they feed on whatever they find, whether it be fresh or spoilt fish, freshly killed or rotting meat, worms, shellfish. Once they spot the carcass of an animal floating on the sea or stranded on the beach, they soon pick it clean, thus earning the title "Sea Vultures" which Buffon conferred on them. When one of them discovers the carcass of a dead whale, it immediately alerts its fellows who promptly and noisily descend on their meal. They are voracious eaters. Yet their stomach quickly digests the carrion which is their preferred food.

There are times when gulls have to go without food for days at a time, for instance, during storms; but they are well equipped to survive the hunger which they have to undure when this happens.

The nest in which they lay their two or three eggs is nothing more than a hole dug in the sand or a cleft rock.

These birds can be domesticated fairly easily. But their meat has a very unpleasant leathery quality which makes it hard to eat. In order to make them edible, sailors hang them up by the feet, and leave them in the open for two or three nights. In this way, they lose some of their smell. (B/C)

(A)

(B)

(C)

Terns

Terns look rather like swallows, with their long, pointed wings and forked tail. And they are quite as restless. Their feet are very short, and their bill, which is straight, slender and sharp, is at least as long as their head. They climb to high altitudes and then swoop down on the prey that their keen eyesight has detected on the surface of the water. Sometimes, they can be seen skimming over the waves at great speed, snapping up any fish which might emerge, however briefly, from the waves. They fly constantly while out at sea and seldom swim at all. When they need to rest, they fly to some isolated rock, far from land.

Terns are gregarious creatures. They feel such a strong bond to their fellow-terns that when one of them has been hit by pellets from a shotgun, the others gather around, and will not abandon the wounded bird until it is clear that it cannot be saved.

In flight, these birds emit a piercing cry which can add up to quite a deafening din when several birds are together. Their call is louder just before they set off on some long journey, and is exceptionally ear-splitting and discordant at nesting-time.

Again like the swallows, terns arrive in spring, on the sea coast. Some remain there all summer, while others move inland to lakes or large ponds, where they feed on any animal matter they can find, fresh or rotting, as well as fish, molluscs and insects. They stay up late into the evening, and search for food even after the sun has gone down.

Terns nest at the edge of the sea and the lakes, in marshes and in wooded areas near estuaries, placing their nests so close together that the brooding females can touch each other. Their two or three eggs are laid on rocks, or in small holes in the ground which look totally unlike a nest. Their eggs are such a delicacy that they are sold in large quantities in some countries.

Terns can be found all over both Old and New World continents, even at the tip of South America and in the islands of the Pacific.

Auks: Razorbills

These birds, found only in cold climates, have a thick, oily layer of fat all over their bodies.

They hardly ever leave the coasts, coming ashore only at breeding time, unless strong winds and heavy breakers force them away from their favorite element. They have an upright stance on land, sitting on their rump. They hold their head very high and their neck outstretched, while their short, stumpy wings look just like arms. When seen from a distance, as they walk along the top of some rocky prominence, they could easily be taken for a squad of soldiers in marching order.

At certain times of the year, the razorbills assemble on the shore and spend one or two days in what looks like a solemn debate. Once an agreement is reached on the subject of the meeting, they set to work. On a fairly uninterrupted piece of land, of about five acres, they mark out a square, with one side, parallel to the water, left open for entry and exit. They then collect any stones lying on the ground within their square, stack them outside the lines, and use them to build a series of low walls punctuated by occasional door-like openings. Sentries are posted at these doors during the night. The birds then divide the land into squares big enough to accommodate a certain number of nests, leaving a path between each square: a very professional job.

A strange association sometimes takes place around breeding-time, as the razorbills, a cross between fish and bird, are joined by an essentially air-borne creature, the albatross; the two species nest side by side. Though vastly different in appearance and habits, these birds live on the best terms with each other, and both feel equally at home. Occasionally, the razorbill will raid its neighbor's nest, but that is all.

Other seabirds land from time to time to seek hospitality from this miniature animal republic. With their host's permission, they nest in the vacant squares.

The female of the razorbill lays a single egg, and rarely leaves it. The male replaces her when she goes off to look for food.

Razorbills abound in such huge numbers in the arctic seas, that Captain Mood collected 100,000 of their eggs in one trip.

(A)

Penguins

Penguins have a shape and structure which is so similar to those of the Razorbill that many travellers have been unable to tell one from the other. Yet the differences between the two species are quite appreciable.

The wings of the penguin are really atrophied and unsuitable for flight: they are a sort of flattened, very short stump. These flappers, which lack real feathers, are covered with a light layer of down, rather like hair, and which could well be thought to be scales.

Situated somewhere between fish and birds, penguins are skilful swimmers and incomparably good divers. For these reasons, hunting them can be very difficult. Their skin is even tough enough to withstand lead shot.

Everything about these birds has been designed for life in the water. Their feet are set far back, which makes them cumbersome on land; the penguins in any case come to land only when they want to breed, which they do in early October, assembled in huge flocks. They merely dig in the sand a hole big enough to accommodate the female's two eggs; as a matter of fact, she frequently lays only one egg. In spite of the low fertility of these birds, they abound in prodigious numbers in the far north.

Penguins do not fear man; they stand their ground, resisting attack with jabs from their bill. If pursued, they pretend to go off to the side, and then quickly turn back and attack their pursuer's feet. Sometimes, as one traveller reports, "they look at you with their head tilted first to one side and then to the other, as if they were making fun of you." Their cry sounds just like the braying of a donkey.

(B)

Puffin

Puffins have the same habits as all birds whose wings are inadequate for flight and whose feet are little use for walking: the water is clearly their only true element. Puffins come ashore only when they need to rest, to build their nests or to escape from stormy weather. They are migrants, leaving the polar regions in autumn and returning in spring. These long flights are sometimes fatal as whole flocks of these birds have been known to get blown to their deaths against the rocks by gale-force winds.

(C)

(A)

(C)

(B)

Auks: Guillemots

These birds have a long straight bill which is convex on top, and angular beneath; the tip of each mandible is slightly curved and indented. Their short, squat legs are set far back. The three forward-facing toes are all joined by the same membrane, and have curved, pointed talons, with virtually no big toe. They have narrow wings and a short tail.

As the shape of their legs makes standing up very difficult, these birds come ashore only to breed or to avoid bad weather; and when they do, they take care to choose the tops of steep rocks, from which they can return quickly to the sea if disturbed. Their nests can be seen covering the tops of such crags, above a sheer drop into the sea. This is the best place to look for guillemots. Each female lays a single, very large egg.

Guillemots feed on fish, insects and crustaceans. They live in the far north but emigrate to the temperate zone when ice begins to cover the sea. They do not fly very well, as their wings are rather short; yet they have to fly on their migrations, so they fly very close to the water, never climbing to any altitude.

The *Guillemot (Uria aalge)* has a bill which is longer than its head; the *Black Guillemot (Cepphus grylle)* is a much smaller bird. (A)

Divers

The distinctive trait of the divers is their bill: it is longer than their head, straight, tough, almost cylindrical, with a sharp point, slightly flattened at the sides, and has the upper mandible longer than the lower. Instead of having separate membranes, their toes are joined by a single membrane.

These birds are fearless swimmers; they dive so fast that they can dodge shotgun pellets before they leave the gun, merely on seeing the flash.

They spend the whole of their life in the water. On land, they have to maintain an uncomfortable upright position which sometimes causes them to fall flat on their bellies, unable to stand up again without immense difficulties. They come ashore only to nest, and then they choose deserted headlands and islets; they lay 12 oblong eggs, usually colored to varying degrees.

Their diet consists of fish, often caught on the bottom, fish fry, water insects and even vegetable matter. Their meat is leathery and quite unpleasant to the taste.

The *Great Northern Diver*, known in the United States as the *Common Loon*, is a beautiful bird with checkered black and white upper parts, with a white belly and collar. The head is black but has some green in it. When this bird has young, instead of diving to escape from its enemies, it attacks with its bill.

The *Black-Throated Diver*, known in US as *Pacific Loon*, has a black bill and throat, with a light grey patch at the top of its head; the front and sides of its neck are white with black streaks; its back and rump are black; its upper wing-surfaces are covered with white patches, and all its underparts are pure white. It occurs in the lakes of Siberia, Island and Greenland, in Hudson Bay and sometimes in the Orkneys. The Lapps make hats out of its feathers. In Norway, it is considered wrong to kill these birds because their various calls apparently enable one to tell whether the weather is going to be fine or rainy.

The *Red-Throated Diver*, known in the United States as the *Red-Throated Loon*, has a mousey-grey color on its throat and the sides of its head and neck; the crown and the back of the neck are striped black and white, and the front of the neck is a bright maroon red; the breast and underparts are white, and the upper parts are black with white patches. This bird lives in the arctic seas. (B)

Grebes

Grebes have a small head, a long neck, an oval body; their legs are very close to the abdomen, with short tarsi. A membrane joins all the forward-facing toes at their base, but the rest of the length of each toe is lobed.

These birds prefer a fresh water habitat, but they also live on the sea. They feed on small fish, worms, molluscs, insects and aquatic plants. Besides being admirable divers and swimmers, they are also very good fliers; yet they tend not to use their wings a great deal, taking to the air only when pursued or migrating. In autumn, they migrate to inland lakes, and in spring they migrate to choose a suitable place in which to lay their eggs.

Grebes usually nest in a clump of reeds at the water's edge. On the outside, their nest is made of roughly woven long grass, and, on the inside, of neatly arranged leaves, grass, etc. They lay between 3 and 7 eggs.

When on land, they move about with a kind of crawling motion, since they cannot walk; they have to sit bolt upright, resting on their rump, their toes and tarsi stretched out to the side. Yet their awkward gait on land is certainly offset by their supreme elegance in the water.

Grebes may be found in both Eastern and Western Hemispheres. (C)

(C)

(B)

(A)

3 – Tropical Birds

Maipouri

These South American birds are found from Mexico to Guyana. They live in wooded country, usually near water. They fly in small groups, often engaging in fierce battles with birds of their own species. They tend to stay well away from human settlements, and strongly resist attempts to train them to do anything.

(A)

Parrots

Parrots have large, strong, rounded bills; the very curved and pointed upper mandible juts out over the lower mandible, which is deeply indented. Their tongue, which is thick, fleshy and mobile, ends in a cluster of nerve papillae or a tip made of cartilage. Their tarsi are very short; their feet are so highly developed as to be almost like hands, capable of picking up, carrying and manipulating any small object. Their toes end in strong, hooked claws, which account for their remarkable climbing ability.

With a few rare exceptions, parrots do not walk easily; it is such an effort for them to drag themselves along the ground, that they very rarely land at all, and only for some very serious reason. They can find all they need to live up in the trees. They are, moreover, poorly equipped for flight, and for fairly obvious reasons: as they live in thick forests, they are never required to travel more than a short distance, from one tree to another. It is true that some species, particularly the smaller ones, can fly quite fast; but they are an exception. On the whole, parrots seem quite happy with a sedentary life within one area.

Parrots are sociable birds; they gather in groups of variable size, sending their deafening call echoing throughout the forest. There are some species which are so strongly drawn to the communal life that naturalists have called them the "inseparables".

At mating time, the birds withdraw in pairs. Male and female show a great devotion for each other. The eggs are laid in hollow trees or cleft rocks. After 20 days of incubation, the young hatch, looking quite naked, and do not acquire all their feathers until three months later. The parents show the greatest solicitude for their young, and become threatening if anyone tries to come too close.

Their diet consists mainly of fruit; they love to devour the kernels of the fruit of the palm-tree, the banana-tree, the coffee-shrub and the guava-tree. Perched on one foot, they use the other to lift the food to their mouth and turn it over until it can be broken easily. Once they have taken the kernel out, they strip it of its wrappings, and swallow it in small pieces. They frequently descend on plantations near their nesting-areas, causing great havoc.

In the domestic state, they will eat almost anything. In addition to seeds, they eat bread and meat, whether cooked or raw; they simply love gnawing on bones. Sugar is a favorite of theirs, also. Bitter almonds and parsley are known to make them violently ill.

Parrots drink and bathe a great deal; in summer they seem to derive a sensuous delight from plunging into the water. In captivity, they readily become habitual wine-drinkers, becoming jolly and talkative under its influence, just like man.

They have a very special way of climbing, without the jerky motions characteristic of other birds of the same order. They move slowly and methodically, supporting their weight on both bill and feet.

Like most tropical birds, parrots are adorned with the most beautiful colors, mainly green, and then red, blue and yellow, in that order. Many of them have highly developed tails.

In spite of their noisy cackling, parrots have become one of man's favorite birds, thanks to their remarkable talent for imitation. They learn to repeat, with great facility, any words which they are taught, or which they may have happened to hear. They sometimes mimic animal calls, musical instruments, etc., with startling fidelity. With their capacity for sudden comical

(B/C)

(C)

(B)

(A)

(B)

utterances, these birds have become a very agreeable companion for man.

Parrots were very common in ancient Rome, sometimes even appearing at Imperial banquets.

In the seventeenth century, a cardinal paid a hundred golden écus for a parrot because it could recite the Apostle's Creed correctly. A Monsieur de la Borde once reported seeing one apparently pleading with a ship's bursar: it was actually reciting the Prayer for Sailors. Willoughby describes a parrot who, on hearing the order "Laugh, parrot!" immediately burst out laughing, only to exclaim, a moment later "Some boob is trying to make me laugh!" A glassware merchant had one who used to say, with feigned anger in its voice, "You dope! What can you expect?" whenever one of the merchant's assistants accidentally dropped something in the store.

The Parrot family consists of four main groups: the *Macaws*, the *Parakeets*, the *Parrots*, and the *Cockatoos*.

The *Macaws*, biggest of all the parrots, have distinctive bare cheeks, and a long tiered tail. These residents of South America have very brilliant plumage. The main species are the *Scarlet Macaw*, the *Blue and Yellow Macaw*, the *Military Green Macaw* and the *Black Macaw*. Their name is thought to have something to do with their ear-splitting shriek. These friendly birds are easily tamed and can safely be given their freedom, as they will always return home to roost. They like to receive attention and caresses from people they know, but are reserved with strangers. Some macaws are noted for their hostility to children, a feeling which is no doubt caused by their jealousy on seeing the children granted so much attention by their mistress.

Macaws have rather poor powers of imitation: they have much difficulty remembering any words at all, and pronounce them badly.

Parakeets are much smaller than macaws; like them, they have a long tail, built in tiers, but their cheeks are covered in feathers. The *Thick-Billed Parrot* is a variety of parakeet which has the area around the eyes bare of feathers, just like the macaws.

Parakeets are highly esteemed for their vivacity, their gentle manners and the ease with which they learn to talk. Their plumage is usually a uniform green, sometimes mixed with red or blue. They are found in South America.   (A)

(A)

(A)

(A)

Parrots can be distinguished from other groups of the same family by their short, square tail. Their cheeks are feathered like those of the parakeet; their size may be anywhere between that of parakeets and macaws. They are prized for their powers of memory and their ability to repeat anything they hear.

Depending on their size and the main color of their plumage, parrots are divided into several sections. The first of these comprises a single species with grey as the dominant tone, the *Grey Parrot*, native to the West Coast of Africa. Next come the species with mainly green plumage. The *Lories* are parrots with a predominantly red color; they live in the Molucca Islands and New Guinea. The Parrotlets are the smallest birds in the group; the shade of their plumage varies according to the climate. They can be found in South America and South Africa and in the islands of Oceania.

Cockatoos have a shorter tail, feathered cheeks and a white, yellow or red crest on top of their head, which they can move up and down freely. They are the biggest parrots of the Old World; they live in the Indies and the islands of Oceania. They are pretty, graceful, affectionate birds, but cannot talk.

(A/B/C)

(C)

(A)

(B)

(C)

Fruit Pigeon

This bird can be found in South Asia, Philippines and Oceania. Its most distinctive trait is its plumage: a gold and green mantle, sometimes purple and maroon, and a tawny or pink grey breast and throat. There are more than sixty known varieties.

(A)

Cotinga

This bird inhabits the warmer parts of the New World. Its bill points downwards, like the flycatcher's, but is shorter. The colors in its plumage are remarkable, and vary according to the season. It is a very wild bird, with a rather sorrowful air, and prefers to live deep in the thickest forests. On the rare occasions when it can be heard at all, it has a distinctly mournful quality to its voice.

(B)

Waxbill

This tiny bird is found in the wild state in Africa, Asia and Australia. It has a conical bill, short legs and a long tail; its brilliant plumage often has patches and streaks of different colors.

In many parts of West Africa, large numbers of waxbills are netted each year, shipped live to Europe and sold as pets. Unfortunately, cold climates are very bad for them; they often huddle together in their cages, trying to keep warm.

(C)

(B)

(A)

(C)

Hummingbirds

There are two sorts of hummingbirds: those with a straight bill and those with an arched bill. All have the same diminutive stature, the same brilliance and the same habits.

Hummingbirds are the most delightful of all winged creatures. Nature seems to have lavished all its gifts on them: gracefulness, elegance, speed, a splendid coat and great courage. They are truly tiny elves, carrying the flashing colors of rubies, topaz, sapphire and emeralds, as they flutter from flower to flower, surrounded by the rich vegetation of the tropics. Their body is so light, their flight so rapid and their size so minute that, with some species, the eye cannot follow the beats of their finely tapering wings on account of their sheer speed. When they are hovering, they seem completely motionless; it is as if they were suspended on invisible wires.

Specially built for an airborne life, they are in constant motion, looking for food in the cups of flowers. They also drink the juice and nectar of the flowers, but this is not their sole food, as some authors have claimed.

Their tongue, which they use as a trunk, is a microscopic instrument of superb design. It consists of two half-tubes, one lying against the other, able to part and move back together again, like the two arms of a tweezers; it is at all times lined with a sticky saliva, the purpose of which is to catch insects.

Hummingbirds take great care with their appearance. They frequently run their bill through their feathers, to keep them smooth and to preserve their brilliance. They are extraordinarily vivacious and petulant, and there is a warlike side to their character which one would hardly expect from such a frail creature. They attack birds much bigger than themselves, harassing them and pursuing them relentlessly, threatening their eyes until they invariably turn and flee. They even fight among themselves. If two males meet on the cup of a flower, they sometimes leap at each other and then fly off, still fighting noisily, until they are out of sight. The winner then returns to the flower.

The hummingbird's nest is a triumph of architecture; it is about as big as half an apricot or hen's egg. The male brings

(A)

(A)

(A)

(B)

the materials which are then arranged by the female. Woven of interlaced lichens glued together with the bird's saliva, it is lined inside with silky fibers taken from various plants. This cradle hangs from a leaf, from a small branch or even from a piece of straw near the roof of a native hut. In it the mother lays a pair of white eggs, about the same size as a pair of peas, twice a year.

The young break out of their shell after six days of incubation, and can fly a week later. The mating pair caress each other abundantly throughout the whole mating season.

Hummingbirds are vividly portrayed in the different languages. The French name Oiseaux-Mouches has to do with their small size; the English term depicts their humming noise; the Creoles of the Antilles and Cayenne call them Murmures, Bourdons (humming) and Frou-frou; in Spanish they are known by the name Picaflores, or flower-peckers, whereas in Brazil they call them Chupaflores, or flower-suckers; lastly, for the Indians, they are Sun-Hair or Sun-Beams.

Hummingbirds are hunted a great deal, not for their meat, of which there is very little, but for their feathers. The Mexicans and Peruvians used to use them to make richly colored overcoats and dazzling little pictures.

It is very difficult to keep them for long in captivity, not because they are not loving and gentle, but because their lively yet delicate nature does not readily accept the confined horizon of a cage. They die after a few months.

Their most dreaded enemies include a monstrous spider, the bird-eating spider, which spins its web around their nest, and devours the eggs or the young, in the absence of the parents who are, themselves, attacked from time to time.

They are resident throughout South America, and are also found in parts of North America; but it is in Brazil and Guyana that they are most abundant. Of the 150-odd known species, the most remarkable are the Crimson-Topaz, the Purple-Carib, the Crested Thorntail, the Frilled Coquette, the Black-Throated Mango, the Giant Humming-Bird, which can be as big as a swallow, the Cuban Bee-Hummingbird, which is no bigger than a bee; also the Tufted Coquette, the Ruby-Topaz Hummingbird, the Amethyst Hillstar, the Sungem, Xantus' the Red-Tailed Comet, and the Racket-Tailed Hummingbird, so called because of its tail which has two fine, long streamers, each ending in a fishtail. (A)

(A)

(A)

(A)

(A)

Turacos

The Turaco, a native of West Africa, is an elegant bird, with its long body and long, slender tail. Its eyes are big and brilliant and are surrounded by a narrow rim of bare skin covered with tiny fleshy knobs. Its tongue is narrow, triangular and pointed, thought full at its base.

According to one naturalist, these birds have a remarkable property: A dozen or so of their wing-feathers, of a brilliant violet-purple color, lose that color on living birds when they have got wet in the rain; if you were to touch them in that condition, your fingers would instantly become stained. However, when they dry, the feathers resume their original brilliance. This does not happen with the feathers of dead birds.

The *Kuyana Turaco*, from the Cape of Good Hope, is quite tame. It is so trusting that it voluntarily follows man, hopping from branch to branch perhaps out of curiosity, and emitting a call which sounds rather like *Cor* pronounced deep in the throat, with repeated emphasis. This cry, which is an expression of pleasure, is accompanied by a graceful lowering and raising of the tail, which it sometimes opens up in a fan shape, while it flaps its magnificent wings, displaying their superb pink coloring. Besides this call, which is also the bird's mating call, it also makes another sound, very like *Corow*, repeated 8 or 10 times, deep in the throat, like a rolled "r". Its cry of alarm consists of a series of strident trumpet-like sounds. (A)

Imperial Pigeon

This bird is really a kind of tropical pigeon; its principal range is Fiji and the Sunda Isles. (B)

(A)

(B)

Toucans

Toucans are distinguished by their inordinately large bill, with its curved tip, serrated edges and a prominent ridge running down the middle of the upper mandible. However, it is not as heavy as one might think; being made of a spongey tissue, honeycombed with air-filled cells, it is no encumbrance to the bird. It is also rather weak—to weak to break things, or even to crush fruit, in spite of its formidable appearance. It cannot be used to cut into tree-bark, either.

Inside this astonishing bill, there lies an even more curious tongue. It is extraordinarily narrow and runs the full length of the bill; each side of it is covered with closely-packed barbs, very like the barbs of a feather. The precise functions of this most unusual feather-like instrument remain a mystery to this day. In fact, the tongue was the one feature which most impressed the naturalists in Brazil, where these birds abound, since the word *toucan*, which they chose to designate them, is a Brazilian dialect word meaning *feather*.

Toucans congregate in groups of between six and ten, in areas with plenty of palm-trees, as their diet consists mainly of the fruit of this tree, and also other fruits and insects. When about to eat a fruit or an insect, they flip it in the air with the tip of their bill and swallow it whole. If it is too big, they cannot cut it up and therefore have to discard it and try something else. Toucans are seldom seen on the ground; although their flight is slow and cumbersome, they nevertheless spend much time moving about from the top of one tall tree to another. Their voice is a kind of hissing noise, which they make quite often. It is very hard to get anywhere near one of these birds. During the breeding season they attack weaker birds, drive them from their nests and gobble up the eggs with the unhatched chicks inside.

They nest in holes previously dug in trees by magpies, laying two eggs. All toucans have a very bright plumage.

The toucan family is divided into the Toucans proper and the Aracaris. These latter are smaller, have a more solid bill and a longer tail. A Toucan with a particularly wide range is the Toco Toucan, which is found throughout Guiana, Brazil, Paraguay, Bolivia and Northern Argentina.

(A/B)

Cocks-of-the-Rock

These birds wear a double, vertical crest of fan-like feathers. They live mainly in the great South American forests, in small flocks of 3 to 8 birds of the same sex. They feed on berries and stone-fruits. Their flight is rather heavy. They nest high up on steep cliff-faces, near mountain waterfalls.

The *Orange Cock-of-the-Rock* is about the size of a Wood Pigeon, and is a beautiful bird. The *Peruvian Cock-of-the-Rock* was long regarded as a sub-species of the preceding species. It differs from it, however, by its larger size, and its colors.

(C)

(B)

(A)

(A)

(C)

Jacamars

The Jacamars live in equatorial America. They have a long, pointed bill, short tarsi, short or blunt wings, and, depending on the species, three or four toes.

Little is known about their habits, apart from the fact that they live alone or in pairs, they are heavy and tend not to move around very much, hardly ever leaving their chosen place of residence. The different species tend to choose a variety of areas in which to live: some like thick forests, others prefer open country, while yet others seem happiest in damp surroundings. All, however, eat insects. (A)

Trogons

The bright colors and irridescent tones of the plumage of this South American bird seem to be the only gift which nature has bestowed on it. Its thick neck and short feet are quite ungainly. This sullen bird lives deep in the heart of the forests, alone or in pairs. Perching on the lower branches of the trees, it stays still and silent all day long. But the mating season, of which there are several each year, rouses it from its torpor; it is then that its melancholy cries can be heard.

Both sexes help build a rather plain nest, dug out of the rotting bark of some old tree. Three or four eggs are laid; the young are born naked, but soon acquire a coat of down. As soon as they can feed themselves, they leave the nest, doubtless animated by the desire for solitude which is characteristic of the species. These birds are hunted intensively for their beautiful feathers, which fetch a high price, and also as delicious game birds. (B)

Hornbills

The remarkable thing about the Hornbills is the huge size of their bill which is serrated and, in some species, has a prominent and bulky shape built on top of it. This bill is nonetheless light, being cellular in structure like the bill of a toucan.

Hornbills have something of the awkward gait of the rook; they are poor walkers and flyers and spend almost all their time perched in tall trees. Large flocks of them live in the forests of the warm regions of the Eastern Hemisphere, mainly in Africa, East Indies and the Oceanian archipelago, nesting in the hollows of trees. They can eat anything and, although their diet consists mainly of fruit, seeds and insects, they also feed on live prey or carrion. In the Indies, they are domesticated because they help kill off rats and mice in human dwellings. Their plumage is black or grey, with occasional white patches. Hornbill meat is a delicacy, particularly when they have been feeding on aromatic seeds.

A great number of species are known, widely differing in size. The most noteworthy is the Rhinoceros Hornbill, so called because its bill is capped with a sort of huge helmet, very much like the horn of a rhinoceros. This bird lives in the East Indies. (C)

(B)

(A)

(C)

(B)

Sugar-Birds

Sugar-Birds are extraordinarily fond of sugar; the natives of Madagascar call them *Soui-Mangas*, or sugar-eaters. They live in Southern Africa and the East Indies. Like the hummingbirds, they are brilliantly adorned and extremely vivacious; they also share with them the habit of poking their bifidated and extendable tongue deep into the cups of flowers. Their brightly colored feathers are worn only during the mating season.

Honey-Bird

This bird lives in South America, from Mexico to Bolivia and the Antilles. It is small, though lavishly arrayed: the male is ultramarine blue and the female a bright green. Except during the nesting season, they assemble in small bands and go from tree to tree, nimbly darting among the branches looking for fruit or small insects. In addition to a brief, high-pitched call, they also have quite a harmonious chirping song. (A)

Gold-fronted Fruit-sucker

This bird, a native of India and Ceylon, lives in the treetops, and usually moves about only in flocks. It has a remarkable song, which sounds rather like a flute, and imitates the songs of other birds with astonishing ease. (B)

Jacanas

Jacanas have a straight bill, of only moderate length, wings armed with sharp spurs, and toes equipped with long, pointed talons, the big-toe claw being longer than the toe itself. These birds inhabit Africa, Asia and South America. In Brazil, they are called Surgeon-Birds, on account of the talon of their big toe which looks like a lancet. They inhabit swamps, lagoons and the areas around ponds. Though they can walk effortlessly across the broad leaves of water-plants, they swim very imperfectly; some naturalists contend that they cannot swim at all. They can fly quite fast, but never very high.

Jacanas live in pairs. They are very wild; anyone desiring to approach them must do so with great care and usually has to resort to subterfuge. Their strong fighting spirit often leads them into fights with other birds, in which they put their spurs to good use. They will defend their young even against man, not hesitating to sacrifice their lives in the attempt.

The male and female have a such profound attachment for each other that they remain together for ever. They nest among the water grasses, laying 4 or 5 eggs; they brood only at night, as the high daytime temperatures of the regions in which they live are ample to perform this task for them during the day. The young leave the nest as soon as they are born and follow their parents.

The Common Jacana is black with a rust-colored mantle and green flight feathers on the wings. (C)

(A)

(C)

(B)

Birds of Paradise

Birds of Paradise have a straight, tough and rather short bill; their nostrils are covered with velvet-like feathers. The brilliance and variety of their plumage makes them at least the equal of the hummingbird; they also have some very distinctive long feathers, situated in various places, which are their most striking ornament. These sometimes take the form of light, graceful tufts, which trail along at their sides; sometimes they are fine, light feathers adorning the side of the head or the tail.

Birds of Paradise live in a very restricted habitat: they are found only in New Guinea, north of Australia. They live in dense forests, feeding on fruit and insects, but according a special preference to nutmeg. Although some of them are solitary, most Birds of Paradise assemble in large flocks and travel when the monsoon comes. They have a fast, agile flight, rather like swallows. Their great ease of maneuver in the air is due mainly to their long side-feathers; yet those very feathers make it impossible for them to move at all when they are caught by a sudden strong tail-wind. When this happens, they get so tangled up that the poor birds simply plummet to earth.

As soon as they became known in Europe, the Birds of Paradise were the subjects of all kinds of extraordinary fables, which were either imported with them from their native land or dreamed up by the naturalists of the day. It was claimed that

they had no feet, that they hung from trees by their long feathers, that they slept, bred and hatched in mid-air, that the female laid her eggs in a cavity on the back of the male, and even that they spent the mating-season in paradise!

The natives of Papua hunt birds of paradise intensively for their feathers. In days gone by, the Rajahs used to adorn their headwear and their swords with those feathers.

According to the arrangement of their feathers, birds of paradise are divided into the following sections: true *Birds of Paradise, Manucodes, Six-Wired Birds of Paradise, Lophorinae* and *Diphyllodes.*

The most remarkable among these species are the *Emerald Bird of Paradise,* whose breast and neck are a brilliant emerald green, and whose sides are concealed under bushy clusters of plumage of a yellowish color; the *Red Bird of Paradise,* whose long feathers are a fine vermilion and whose throat is a magnificent gilded green; the *Crinkle-Collared Manucode;* the *Superb Bird of Paradise (Lophorina Superba);* the *Arfak Six-Wired Bird-of-Paradise,* so named because of the three wire-like plumes, with tips shaped like paddles, which adorn either side of its head; and, lastly, the *Magnificent Bird of Paradise (Diphyllodes Magnificus).*

4 – Miscellaneous Families of Birds

Woodpeckers

The distinctive features of the birds of the Woodpecker family are their long, conical, pointed bill, and a very extendable tongue.

Woodpeckers are complete masters of the art of climbing. They stretch out their toes over the tree-trunk and hang on by digging in their hooked talons; they then move on with a sudden jump and repeat the process. This technique is made easier for them by the arrangement of their stiff tail feathers, which press against the tree, thus providing support for the bird. This feature of theirs enables them to move freely over the trunks of the trees, both vertically and horizontally.

Woodpeckers are shy, fearful birds. They live alone in the middle, or at the edge of large forests, feeding on insects and insect grubs. They find their food in the trunks and cracks of the trees, using a tongue which is marvellously adapted to this kind of exploration. It is very long and has a special mechanism which enables it to reach objects more than two inches outside the bill; its tip is made of a horny substance, equipped with tiny hooks; moreover, it is coated with a sticky fluid, secreted by two large glands, which serves to trap unwary insects. When a woodpacker darts this tongue into a chink it has found, it always comes back out fairly well laden with insects. Should the use of this device fail to dislodge an insect, the bird then hammers away with its bill to break through the bark and seize its prey. Sometimes it uses its bill to check for the tell-tale echoing sound which reveals the presence of a cavity in which insects might be hiding. If the trunk does sound hollow, it examines every inch of it carefully to find a way into the cavity, using its tongue, and, if necessary, enlarging the aperture with its bill, until no corner is left unsearched.

Food is not the only reason why woodpeckers make holes in tree trunks: they also nest in them. While some species are content to use any natural crack they might find, other species prefer to carve out their own nest, according to their own tastes. This is what they are doing when one sees them inspecting softwood trees, like beech or aspen; they are looking for the ones with internal flaws. Having chosen a site, the male and the female take it in turns to hammer their way into the trunk, stopping only when they have reached the rotten part. The entrance which results from their pecking is usually so deep and so oblique that it is quite dark inside, which no doubt offers them some security against their natural enemies among the small mammals, particularly against the rodents. The female lays her eggs on a bed of moss or tiny, decaying wood fragments.

On the whole, woodpeckers have no voice or at best an unpleasant call. During the mating season, they often use a language peculiar to them alone, and which is loud and resonant enough to attract all woodpeckers within their area: they bang on the trunks of dead trees with their bills.

There are many species of woodpeckers, to be found throughout both New and Old World continents. The main species are the *Black Woodpecker*, the *Spotted Woodpecker* and the *Green Woodpecker*. (A)

Swift

Swifts have tiny feet, quite useless for walking. This is to be expected, as they spend their entire life in the air, landing on the ground only by accident and then having the greatest difficulties in taking off again. They know only two states: either violent motion or total rest. The only intermediate stage for them, between agitated flight and the dark seclusion of their nest, is when they crawl along walls, cliffs or tree trunks to which they cling with their bill and their talon-like toes. Swifts usually fly straight into their nest without stopping, after flying to and fro in front of it a few times first. They build their nests as high as possible, in cracks in old walls. When they have young to feed, they keep a supply of various winged insects in reserve in their ample throat, and return to the nest with food two or three times a day. (B)

Rollers

Rollers have a strong bill and wide-open nostrils; they are very wild birds and live deep in the forests. Although they do come to know people who care for them, they never show much familiarity. Their diet consists of insects, worms and small reptiles, extending to berries, seeds and roots when necessary. They have a remarkably brilliant plumage. They live in Europe, Africa and South Asia. (C)

(B)

(C)

(A)

Bee-eaters

The bill of the bee-eater is long, thin, curved and pointed, and has a cutting edge; its tarsi are very short, its wings long and pointed; its highly developed tail may be evenly shaped, forked, or have a tier-like structure. These are slender, light, noisy birds who, with their sustained, rapid flight, stay aloft for hours and hours at a time. Their name derives from their diet, which consists of hymenoptera, mainly bees and wasps. They catch these insects either while in flight, like the swallows, or by lying in wait outside their nesting-places and picking off all those entering or leaving. At the same, they skilfully manage to avoid being stung. Since they live in large flocks, even at breeding-time, they can quickly decimate those species of hymenoptera which they are interested in throughout a whole area: whereupon, they go somewhere else, looking for food.

The nest on the banks of rivers, at the end of deep tunnels which they dig themselves and which are sometimes as much as 6 ft 6 ins. long.

Bee-eaters usually live in very hot climates; Europe does have one species, the Bee-eater *(Merops apiaster)*, which emigrates regularly each year, arriving in May and leaving in autumn. (A)

Kingfishers

The *Kingfishers* are quite striking birds. Their bill, which is tough, straight and angular, is quite out of proportion with their small size; they have a strong, elongated head, and a rather stout body; their short tarsi are set far back; all of this, plus their richly colored plumage, in which blue predominates, causes them to attract much attention. They live at the water's edge, where they feed, as their name suggests, almost exclusively on fish, which they catch expertly and with much patience. Perching on a dead branch, or on some rock in the middle of the water, the kingfisher waits, motionless, for hours at a time. As soon as it spots its prey on the surface, it closes in quickly, and grips it in its powerful mandibles; then, once the fish has been crushed or beaten to death against a rock or tree-trunk, the kingfisher swallows it, head first. When fish are in short supply, it will feed on water-insects; but this time, it will feed on the wing. It flies in short, powerful bursts, never climbing much above the ground. The shortness of its tarsi make walking very difficult.

Kingfishers are anti-social; they always live alone, except at mating-time, in the spring. They nest on river-banks, in natural crevices or the holes left by water-rats; their nest becomes littered with the left-overs from their meals.

The mother and father take turns at brooding, feeding their young on the results of their fishing-trips. These birds have no song and their meat has an unpleasant taste.

Kingfishers were the Halcyons of ancient times. A body of legend came to be built up about them; for example, they were supposed to be able to indicate the wind after their death, and to dry up any wood on which they landed. Their desiccated bodies were thought to be capable of diverting thunder, imparting beauty and bringing peace and abundance.

Kingfishers are to be found all over the world.

There are many species, particularly in warm regions of Africa and Asia. Europe has one of the prettiest species; it is about the size of a sparrow. (B)

Nightjars

The main features of the nightjars are: a very wide gape, short legs, a versatile big toe; they have the soft, downy plumage and physiognomy of all birds of the night. They are sad, lonely birds who live in pairs, sleep during the day, and come out only at sunset to hunt for the insects of the twilight and night hours. They thus spend the whole night flying; their prey is scooped into their wide-open bill as they fly along and trapped there by a special sticky saliva which covers their palate. Since they feed mainly on moths, dragonflies, crickets, cockchafers, bumble-bees, flies, etc., they are just as valuable as swallows and should be treated with the same respect. However, in autumn, when Nightjars are plump and appetizing, hunters are only too eager to sacrifice them for the pleasures of the table.

The species which is most typical of the true nightjar is the Nightjar *(Caprimulgus europaeus)*, which is about the same size as the blackbird. This bird often flies among grazing herds, and picks off the insects which pester the cattle. The Nightjar can be recognized by the moustaches at the base of its bill, and also by the joint of its middle toe, which is serrated. The bird is thought to use this comb-like object to scratch its head, and to wipe away surplus insects from around its mouth. Nightjars are migrants; they travel slowly, by night. Another curious feature of theirs is the noise made by the air rushing into their open bill as they fly. (C)

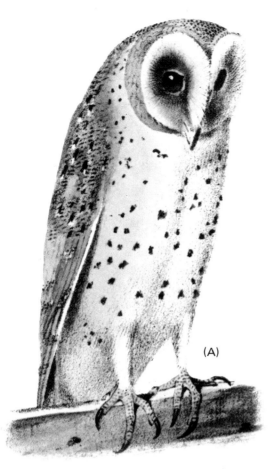

(A)

(D)

(B)

(C)

The Long-Eared Owl is about a foot tall. Its ear-tufts are correspondingly shorter than those of the Eagle Owl. Its wingspan is three feet. Its plumage is predominantly rust-color, mottled with grey and brown. Its bill and talons are blackish, and its eyes are a beautiful yellow. It roosts in cleft rocks or dead trees and old abandoned farmhouses. Sometimes it finds it convenient to take over nests left empty by magpies, ravens or buzzards. Being much less wild than the Eagle Owl, it often comes close to human dwellings. Mice are its favorite and principal food; it can therefore be quite easily lured into a trap if one imitates the cry of a mouse. Its diet also includes moles, field mice, frogs, toads, even young hares and rabbits, and, when all else fails, insects. Its cry is a sort of low moan: clow, cloo, which it makes frequently during the night. If captured young, it can be tamed quite easily. Adults will not eat in captivity, and simply die in their cage.

Scops Owl is noteworthy for its small size, about the same as that of a blackbird, and also for its ear-tufts, which are quite rudimentary, consisting of a single feather. Its plumage, speckled with rust, grey and black, is prettier than that of the above species.

Scops Owls assemble in flocks in autumn and spring to travel to other climates; they leave after the swallows, and arrive at about the same time. Scops Owls are of great value to agriculture, as they destroy large numbers of field mice.

There is a variety known as Scops Asio which may be found in North America, on the banks of the Ohio and the Mississippi. It is a gentle bird, and allows itself to be stroked, when captured, without trying to scratch or bite. Audubon describes how he took one from New York to Philadelphia; he kept the bird in his pocket throughout the journey, and taught it how to eat out of his hand. The Scops Owl made no attempt to escape.

Cuckoos

The Cuckoos have elegant bodies, with a slightly curved bill almost as long as their head, a wide gape, and quite a long, rounded tail. Unlike other birds of the same family, they have long, pointed wings. As for size, they are somewhere between the blackbird and the lark. Their flight is fast and easy, but they cannot withstand even moderately strong headwinds; for this reason, they cannot undertake long journeys without resting.

There are a great number of species which all belong to the Old World. Cuckoos may be found in the whole of Africa, in Southern Asia and certain islands of Oceania. Europe has one species only, known simply as the Cuckoo.

The cuckoo is essentially a traveller. It spends its summers in Europe and its winters in Africa, or in the warm parts of Asia. It arrives in April and leaves in late August or early September. It travels at night, not in large flocks but alone, or, at most, in groups of two or three. It spends much of its time in the thickest part of the woods, but also often flies over the open countryside looking for its food, which consists of insects, particularly caterpillars. It has a voracious appetite, this being due to the large size of its stomach. It is by nature tyrannical and nasty, and will tolerate no rival of its own species in the territory which it has chosen.

The cuckoos are famous for the rather peculiar way they have of raising their young. The females make no nest, do not brood and do not care for their young. They leave their eggs in other bird's nests, for example, in those of the lark, warbler, robin, nightingale, thrush or blackbird; sometimes they lay them in the nests of the magpie, turtle-dove or wood pigeon. They let these strangers hatch their eggs, and feed their offspring until they are fully grown. Several explanations have been offered for this anomaly.

Having left her eggs with their new nursemaid, the female cuckoo comes several times to make sure that they are being well taken care of, and leaves the area only when she is quite sure that this is so.

It can now be seen why the female cuckoo does not perform her own maternal duties. As she lays her eggs at very long intervals, she would find herself brooding several eggs and raising a young one at the same time.

Within a short time after hatching, the young cuckoo uses such strength as it has to get rid of its new mother's real offspring, so that it alone will receive her attention. It slips under the young chicks, loads them on its back, steadying itself with raised wings and, dragging itself to the edge of the nest, it tips them over the side, one by one. In most cases, the mother continues to treat the treacherous adopted child with tenderness, and provide for its needs until it leaves the nest; but sometimes she brings the monster no food at all, and leaves it to die of starvation. (A)

(A)

(A)

(A)

Vultures

These birds have a bare head and neck, with a collar of feathers around the base of their neck; round or oval nostrils; tarsi bare or with feathers covering the upper half; very long middle toe; pointed wings, trailing to the ground. Their flight, although powerful, is slow and heavy, whence their name (Vultur: *volatus tardus*, delayed flight). They love carrion, but have been known to take an occasional meal of fresh meat.

The *Griffon Vulture*, about the size of a goose, is commonest in South and South-East Europe. It nests in the most inaccessible cleft rocks. When sufficiently hungry, it readily attacks live animals.

The *Bearded Vulture* is slightly bigger, with a height of 40–45 inches. It is to be found in Greece, Spain, Egypt and much of Africa, and has also been sighted several times in the Alps. In the fall, it leaves the temperate zone for warmer climates, where it spends the winter.

The *Lappet-Faced Vulture* inhabits the mountains of Africa. It has a fleshy comb extending from its ear down the whole length of its neck. This extremely voracious bird can be as much as 55 inches, with a wingspan of nearly ten feet. It builds its eyrie on steep cliffs, where it is very difficult for anyone to approach. (A)

Condor

Condors live in South America, over a range extending from Patagonia to the Equator. Their preferred habitat is the isolated upper regions of high mountain ranges, quite often near the permanent snow line. They usually come down to the plains only to look for food.

The Condor has a more powerful flight than any other bird; a few slight flaps of its wings seem able to transport it above the clouds in a matter of seconds. Humboldt saw Condors soaring at 22 thousand feet above sea-level. Another naturalist saw one take off high in the Andes and climb until it was completely out of sight against the blue sky above. How they manage to live in such rarified air is still a mystery.

Garcilasso, in his History of the Incas, claims that these birds have been seen devouring children of about 10 or 12 years old, and that two of them would suffice to kill and eat a head of cattle. Travellers in past centuries have said that Condors are strong enough to carry goats, sheep and lamas up to the peaks of the Andes. However, since their big toe is too short to enable them to grip with their talons, such assertions have been questioned; it is possible, though, that they might be true as the Condor is thought to use its awesome bill to flip its living prey onto its back, and fly with it in that position back up to its eyrie. (B)

Buzzard Hawks

Buzzards have long wings, a large head, a stout body, short or mediocre tarsi and a bill which is curved from its base; in other words, they look rather ungainly and heavy. They do not take their prey on the wing; such violent exercise would be too much effort for them. Rather, they prefer to lie in ambush on a tree or a mound of earth, and wait there patiently until their prey comes within reach. They sometimes stay in this position for several hours at a time, completely motionless, with a stupid expression which has become proverbial and which is due as much to their dejected posture as to the weakness of their eyes, since they are very sensitive to sunlight.

Buzzards nest in the higher trees of large forests in either plains or mountains or in the midst of brushwood growing among rocks. At the first sign of frost, they move closer to human dwellings, in the hope of catching some poultry; when really hungry, they do not hesitate to grab them in broad daylight. Their diet consists mainly of small birds, rodents, snakes, insects and, occasionally, cereals. They are easily tamed. A story is told of one buzzard who had established a fine relationship, based on mutual understanding, with a retriever; they even shared their meals together.

Buzzards readily undertake the incubation and raising of young birds.

Mr. Yarrel reports that, in the town of Uxbridge, in England, a domestic buzzard had indicated the need to build a nest; she was given the necessary materials and, when the nest was complete, two hen's eggs were placed under the bird. She sat on them, hatched them and raised the young chicks as if they had been her own. One day, in order to spare the buzzard the trouble of brooding, some freshly hatched chicks were placed in her nest; but she killed them all, because she had not performed any maternal duties towards them.

The main species of buzzard are the Common Buzzard, which is to be found throughout Europe; the Honey Buzzard, which inhabits Eastern Europe; its favorite and principal food is wasps and bees; it also eats wheat and, in the domestic state can manage fruit quite nicely; the Rough-Legged Buzzard, so named because its legs are covered with feathers down to the toes, inhabits Europe, North Africa, Asia and America, where it is known as the American Rough-Legged Hawk. (C)

(A)

(B)

(C)

Falcons

The Falcons (from falx, scythe, for the shape of its talons) are extremely well adapted to the life of a predator, in fact they are the ultimate bird of prey. Their short bill, which is curved from the base, has a very powerful tooth on each side of the upper mandible, fitting into a notch on the lower mandible. Long, pointed wings give it a powerful, fast and agile flight. Its legs are short, ending in sharp, hooked talons. These features, together with the most penetrating vision and enormous strength explain why this bird strikes terror wherever it goes. They feed only on living prey, birds or small mammals, which they often kill with a single blow from their deadly bill; they then fly off, with their victim dangling from their talons, to devour it undisturbed. They hunt only when in flight; they gather in groups when migratory birds are on the move, follow their flocks and exact a daily toll from them. Yet they usually live alone in pairs, nesting in woods, on cliffs, in disued mines or old farmhouses, depending on local circumstances. They lay between 2 and 4 eggs.

The falcon family consists of two groups: the Gyr Falcons, whose tail is longer than their wings, and the Falcons proper, whose wings are as long as, and sometimes longer than, their tail.

The Gyr Falcon is the best proportioned and most vigorous of all the falcons. Though barely 2 feet tall, it is quite as strong as an eagle. Its beautiful brown coloring changes with age to an almost pure white. This bird inhabits the arctic regions, feeding on big birds, particularly on gallinaceans and palmipeds.

Among the Falcons proper, the Peregrine deserves first mention. It is common in the centre and north of Western Europe, and also in the islands of the Mediterranean. It also lives in North America, where it is known as the "chicken eater". Its flight is prodigiously fast. It soars effortlessly to great altitudes; and, having spotted its prey, it streaks down on to it, tears it to pieces and devours it greedily. If the victim is a bird, it first plucks the feathers off with its bill and then swallows it.

The falcon feeds on water birds, pigeons and partridges. When necessary, it includes even larks in its diet, pursuing them, on occasion, right into the birdcatcher's net, from which it is not always able to escape. As Audubon saw on the banks of the Mississippi, it is capable of stooping to the point of eating dead fish, but this is rare. The Peregrine is a brazen bird; sometimes, it seizes freshly killed game before the hunter himself has a chance to arrive on the scene.

The longevity of the Falcon is even greater than that of the eagle. In 1797, one was captured at the Cape of Good Hope wearing a golden collar with an inscription stating that it had belonged to King James I of England, in 1610; it was thus 187 years old, but still very lively, nonetheless.

The other species of falcon differ from those above only because of their smaller size; their prey also tend to be smaller, and include quails, larks, swallows and, sometimes, insects. These species are: the Hobby, to be found in the whole of Europe and also in Africa; this bird stands one foot high; the Merlin, about the same size as a thrush; the Kestrel; the Sparrow Falcon, a resident of India and Sumatra and the smallest of all the birds of prey. In addition, there are several types found only in the Americas.

(A/B)

(B)

(A)

(A)

Goshawks

The birds which make up the species Goshawk are different from the falcons in that they have no teeth on the upper mandible, their tarsi are longer and their wings shorter. Although they fly very fast, they do not fly as high as the falcons. With slight adaptations to climate, they are to be found all over the world. They normally feed on small birds and reptiles, and, exceptionally, on very small mammals. They are divided into Goshawks and Hawks.

There are various species of goshawk, all of them characterized by very strong tarsi. The Common Goshawk hunts by flying close to the ground and bushes which it scrutinizes carefully. Upon spotting a victim it approaches with increased caution, until it is very close. Then it suddenly leaps on its prey, striking before it has time to recover. The Goshawk is as big as the Gyr Falcon; yet, although quite as cunning and skilful, it is less courageous. The Carolina Goshawk and Stanley's Goshawk are to be found in America. (A/B)

Eagles

The species Eagle may be described as follows: the bill is fringed but lacking teeth, with a straight section at the base; elliptical and transversal nostrils; tarsus short, with feathers right down to the toes; long wings, rounded tail.

The ancients used to think of the eagle as the bird of Jupiter and the messenger of the gods because it flies to such great heights.

This is a bird with very highly developed vision. The great strength of the muscles which power its wings account for the speed and duration of its flight.

The eagle's size varies according to the species, but is always impressive. The female of the golden eagle stands 45 inches from the point of her bill to the tip of her feet, with a wingspan of almost 10 feet. The wingspan of the Imperial Eagle is only 6 feet, while that of the Spotted Eagle is 5 feet 6 inches.

The eagle builds its nest in the cracks of the least accessible rocks, on the edge of cliffs, so as to protect its young from attack. The nest itself is really a kind of floor made of sticks laid artlessly side by side and connected by flexible branches, with a layer of leaves, reeds and brushwood on top. Nonetheless, this floor is built so solidly that it can withstand, for many years, the onslaught of the elements, and moreover support, besides the 70—90 pounds weight of the 4 or 5 occupants themselves, the usually large amount of provisions which the eagle accumulates.

The longevity of eagles is quite remarkable, though difficult to determine exactly. One naturalist mentions the case of an eagle which lived in captivity in Vienna for 104 years, as well as a pair of eagles in Forfarshire, in Scotland, which had lived in same eyrie so long that the oldest inhabitants had always remembered them living there.

If caught very young, the eagle can be trained; but he never quite loses a certain wild quality which makes him rather sullen and difficult to get on with. At the age of two or three years, it is already very difficult to tame him, as he rewards those who approach him with a painful jab with his bill. When he is old, he is altogether unmanageable.

The eagle is cosmopolitan: he can be found all over the world. A number of varieties exist, but their lifestyle varies only slightly from one species to another. The Golden Eagle is the biggest of all. The Imperial Eagle may be found in the east and south of Europe, and also in North Africa. (C)

(A)

(B)

(C)

(C)

(C)

Ibis

The bill of the *Ibis* is long, curved downwards, almost square at the base and rounded towards the tip; its head and neck are largely bare; it has four toes, three of them facing forward, all joined at their base by a membrane, with almost the full length of the big toe resting on the ground.

These birds inhabit the warm areas of Africa, Asia and the Americas. Only one species, the *Green Ibis*, is found in Europe. They can be seen, in groups of seven or eight, in damp, swampy country, and on the banks of big rivers, where they feed on the worms, water insects and small molluscs which make up the bulk of their diet. They also eat tender blades of grass which they tear from the ground themselves. They are gentle, peaceful birds, without the sprightliness of movement which is so characteristic of many other wading-birds: they will often stay in the same place for hours at a time, prodding the mud for their food. Like almost all the birds of this family, they emigrate each year, making huge trips from one continent to another. They are monogamous, to such an extent that death alone can break their vows of eternal fidelity. They usually nest in tall trees, occasionally on the ground, and lay two or three whitish eggs, which hatch after 25 to 30 days.

There are between eighteen and twenty varieties of Ibis. The *Sacred Ibis* is about the same size as a hen. Its plumage is white, with black rump and wingtips. Its fame dates back to ancient times, when the Egyptians treated it as an object of worship. They used to raise these birds in the temples, as if they had been a divinity; they were even allowed to run around the streets, to such an extent that, according to Herodotus and Strabo, traffic jams used to ensue. Anyone who killed an ibis, even by accident, was immediately set upon by a frenzied mob and stoned mercilessly. After their death, these birds were collected and embalmed with every care.

The *Green Ibis*, which Herodotus called the *Black Ibis*, has black plumage with green tones on its upper parts. It lives in North Africa and Southern Europe; like the *Sacred Ibis*, it also was venerated by the Egyptians.

The *Scarlet Ibis* is a native of South America, mainly Guyana, where it lives in flocks at the mouths of rivers. Its plumage is bright scarlet throughout, except for its black wingtips; however, the young birds do not acquire this brilliant coat until the age of two years. The young are easily tamed; their meat is delicious. (A)

Spoonbills

The remarkable thing about the spoonbill is the curious shape of its bill, which is flat, wide and rounded at the tip. It is a useful way of catching worms and the small fish which spoonbills find especially delicious, in both water and mud; they also feed on water insects, by holding their bill half-open on the surface of the water, and catching these insects as they fly by. They congregate in small flocks near the sea shore. Two species exist: the *Spoonbill (Platalea leucorodia)*, which is found in various parts of Europe, and which has a pendant horse-tail crest on the back of its neck; and the *Pink Spoonbill*, found in South America, which has some very fine shades of pink in its plumage. Spoonbills can be domesticated quite easily.

(B)

Herons

Herons have a long, pointed and very strong bill, with a very wide gape; parts of their legs are without feathers; their long toes all have sharp claws, including the big toe, the whole of which rests on the ground; their neck is long and slender; the back of their head is covered with long feathers in the shape of a tuft, trailing along their back, whereas the narrow feathers on the front of the head hang, rather like a beard, at the base of its neck.

These semi-nocturnal birds inhabit the shores of lakes, marshes and rivers, feeding on reptiles, frogs and fish. On the whole, they are rather wild by temperament and live alone in the midst of their domain. When searching for prey, they wade until the water is half-way up their legs, and then stand perfectly still, like a statue, for hours on end, with their neck folded against their chest and their head buried between their shoulders. As soon as a fish comes within range, their neck suddenly reaches out, as if powered by a spring, and one jab from their fast-moving bill spears their hapless prey. When fishing is poor, they trample on the mud so as to force the frogs and other animals that may be there out into the open. If necessary, they will attack rats and field mice, and, when all else fails, they are not above an occasional meal of carrion. They are, in any case, able to go for long periods without food.

Most herons are powerful flyers; they usually emigrate in large flocks, the young and the old travelling separately. However, since they are able to adapt to all temperatures, there are some resident species; they can be found over a wide range of areas, throughout the year. (C)

(A)

(B)

(C)

Storks

The bill of the stork is long, straight, pointed, sharp and broad at the base. The big toe is high, and rests on the ground. Its wings are broad and concave, its tail short. They live all over the world; some species emigrate very regularly. They are well-suited to long journeys, because, though a large bird, the stork is actually very light, most of its bones being hollow.

Storks live in wet, flooded areas, on the banks of ponds and rivers. Their diet is exceedingly complex, consisting mainly of reptiles, batrachians and fish. They also eat birds, small mammals, molluscs, worms and insects, including bees. They are not above an occasional meal of carrion or refuse. They have a solemn way of walking, and are rarely seen running; yet they are good flyers, flying with outstretched neck and legs, a posture which, together with their great wings spread wide, makes them look rather like flying crosses. They have no voice, the only noise they make being a cracking sound when they bang their mandibles together. This, their way of expressing both love and anger, can be quite loud. They lay between two and four eggs, their fertility varying inversely with their size. Their lifespan is between 15 and 20 years.

The stork has a very gentle nature and quickly becomes friendly. As a natural enemy of many animals which are a nuisance to man, it has become a useful assistant, rewarded and protected by man since the earliest times. The Dutch and the Germans still regard it as a happy omen if the stork chooses their house to make its nest. Sometimes, they even make it easier for him by laying a packing case or a wheel on the top of the roof; this is the frame of the nest, which the bird will then complete with his own combination of reeds, grass and feathers.

The White Stork stands between 39 and 47 inches tall; its plumage is white, with black fringes along the edges. This is the commonest species in Europe. The Black Stork is slightly smaller; it lives in Eastern Europe, feeding almost exclusively on fish, which it catches with great skill. It is quite a fierce bird, avoiding the society of men and nesting alone in green trees. The Marabou is easily recognized because of its enormous and very strong bill, its bare neck, the lower part of which is equipped with a large pouch, rather like a fat sausage.

Marabous live in the Indies and Senegal, feeding on reptiles and refuse and thereby winning the goodwill of the people. In the large towns of India, they are every bit as domesticated as the dog in western countries; they scour the streets for refuse thus helping keep them clean. At meal-time they invariably line up outside the barracks waiting for any leftovers the soldiers might toss to them; they are such gluttons that they are capable of swallowing enormous bones whole.

There are several types of bird which are closely related to the stork, only with slight differences in the bill. We shall just name them and state where they live. They are the Jabiru, from South America; the Umbrette, or Hammer-head, from Senegal; the Dromas, which is found on the shores of the Black Sea and in Senegal; and the Tantalus, or Wood-stork, which lives in the warmer regions of both Eastern and Western Hemispheres. (A)

Flamingoes

The Flamingo is one of the most curious waders. Even the wildest imagination could hardly dream up anything so odd as the body of this bird. It has endless legs supporting a tiny body; a neck to match; a bill which is higher than it is wide, with a sharp curve, as if it was broken in the middle; smallish wings; a short tail: broad-webbed feet, a short and raised big toe, and a splendid pink color, becoming bright red on the back and wings, complete the picture of this most unusual physiognomy.

It was this bright wing-coloring that led the ancients to call this bird phenicopter (wings of fire); we have expressed the same idea by the word Flamingo, which, while less pretty than the Greek word, derives from exactly that notion.

Flamingoes live along the shores of lakes and ponds, and only rarely on the seashore or on the banks of rivers. They feed on worms, molluscs and fish eggs which they catch in the following manner: they put their neck and head in such a position that the upper mandible of their bill will be underneath; they then stir up the mud in all directions and find an abundance of food. They also use their feet in order to trample on the mud and uncover the tiny animals they need. They are very sociable and live in flocks, under a strict discipline. When they are fishing, they stand in long and regular straight lines, and post a sentry to warn them of danger. As soon as there is any cause for alarm, he emits a noisy cry, rather like a trumpet; and the entire flock immediately flies away in perfect order.

Flamingoes are thus highly distrustful, but only of man, it seems; the sight of animals is not enough to make them turn and flee. The Flamingo's nest is a cone with a flat top, about 1 foot 9 inches high, made of sun-dried mud. The two rather elongated eggs, matt white in color, are laid in a shallow cavity at the top of this mound. While brooding, the female sits astride this novel kind of throne, her legs dangling at either side. Her young ones can run soon after birth, but are not able to fly until later, when they have all their feathers; only when they are about two years old do they put on the brilliant adult plumage. (B)

(A)

(B)

Swan

The Swan, biggest and most beautiful of the waterbirds, has been admired throughout history. Its noble and elegant proportions, the graceful curve of its neck and the rounded shapes of its body have always inspired the poets: the swan was made the bird of gods and goddesses. The poetic imagination of the Greeks associated all kinds of charming fantasies with its name. It was even thought that, before breathing its last sigh, the swan marked its own passing with a melodious song.

The main strength of the swan lies not in its bill, but in its wings, which are quite a powerful weapon which it puts to good advantage. Its bill is red, with black edges, and its plumage is snow white. It swims with ease and flies perfectly. Its diet consists of plants, rockweed and small water insects; sometimes it even attacks fish.

Swans live in groups in Europe, Asia and the two Americas. In February, they go off in pairs to nest. Between 8 and 10 eggs are laid, of a white and greenish color, and incubation lasts about six weeks. They carry their young on their backs, hide them under their wings for warmth and never leave them in their early youth. The grey down which covers the young swans at first is replaced by adult plumage only in the third year of life.

These exquisite birds occasionally engage in terrible fights among themselves. A fight between swans is a fight to the death, in which the adversaries display both astonishing force and fury, and also quite remarkable patience and skill. Sometimes the fight will last several days, to end only when one of the swans has succeeded in locking its neck around that of its adversary and holding it under water long enough to kill it.

The swan is clearly aware of its beauty and gracefulness for it is constantly preening its plumage. It is perhaps the most coquettish of all birds.

(A)

Geese

Geese are similar to ducks and swans, but spend less of their time in the water. Indeed, they wander quite far from the water, looking for the wet grasslands and marshes where they might find their favorite plants and seeds. They swim infrequently and do not dive at all. They nest on the ground, laying between six and eight eggs which they brood for a little more than a month. As soon as they hatch, the young goslings are able to walk and feed themselves. Geese in general, and male geese in particular, have two molts every year, in June and November.

Flocks of geese looking for their food make a lot of noise, and can be heard a long way off. They have a brassy, trumpet-like call, which they emit frequently. They also make a more continuous, but briefer, clucking noise, which tells the observer that they are coming.

When attacked, geese make a hissing noise like a snake. They are awakened by the slightest sound; they then let out a single, but unanimous, call which seems to alert the whole flock. Some writers have contended, for this reason, that geese are more vigilant than dogs.

Everyone remembers the story of the geese on the Capitol, who saved Rome from a sneak attack by the Gauls. Out of gratitude, the Roman people set aside a certain amount of money each year for the needs of the Capitol geese; on the anniversary of the triumph of these feathered guardians, it became the custom to whip some dogs outside the Capitol, as a punishment for their negligence.

It is hard to understand why geese have always been thought of as the symbol of stupidity. They have penetrating powers of vision, very fine hearing and a sense of smell comparable to that of the raven. Certainly their vigilance is never found wanting.

The flight of wild geese shows a high degree of intelligence. They form up in the shape of the two long sides of a V, or, if there are not many of them, in a single straight line. This arrangement makes it possible for each one of them to follow the rest with least possible fatigue, and also to stay in position. When the bird flying first begins to get tired, it falls back to the rear, and the others, in turn, move up to the front, to guide the flock.

Although they spend very little of their time in the water, wild geese seek out a pond or river every evening, in order to spend the night; in this way, one could say that ducks enter the water only as the geese leave.

(B)

(B)

(A)

(B)

Ducks

The sea-shores and river-banks of the whole world are inhabited by numerous species of duck. No other bird exists in such numbers at the water's edge. Some species are remarkably beautiful and display a dazzling variety of colors. On land, ducks walk with a clumsy, rolling gait, but in the water they are elegant and agile. In their natural element every movement is graceful and effortless. They love dabbling in the mud, where their great appetite finds constant nourishment. They eat a wide range of food: water insects, worms, slugs, snails, small frogs, bread, meat whether fresh or spoiled, and fish, whether living or dead. They are usually so greedy that it is not uncommon to see two of them tugging and quarrelling, for more than an hour, over an eelskin, or other similar scraps, one of them having already begun to swallow one end, while his adversary holds on firmly to the other.

All our breeds of domestic ducks are descended from wild ducks. Their true home is those far-northern countries where man cannot live because of the severe climate. The rivers of Lappland, Greenland and Siberia are literally teeming with them; in May, their nests abound in numbers which it is difficult to imagine.

The flight of ducks is fast, powerful and sustained. They can take off from land or water with a single flap of the wing, and climb vertically above the tops of the tallest trees; then they fly horizontally. They fly at very high altitudes and travel long distances without stopping to rest. They can be seen flying towards their destination in triangular formations and the whistling sound of their wings can be heard from a long way off.

Wild ducks are highly suspicious creatures. When they are about to land somewhere new or move from one pond to another, they fly around, up and down in concentric circles until their reconnaissance of the new landing place is complete.

One must not expect the nests of wild ducks to be particularly elegant. Sometimes they choose a thick clump of reeds, trample over part of it rather haphazardly and simply cut or bend some of the stalks. Yet they line the inside with a fine layer of down. Sometimes their nests are to be found quite far from the water, amongst the heather. It is not unknown for the female to take over the nests of magpies or crows, which these birds have abandoned in the trees.

Audubon describes a remarkable instance of the maternal love of this marsh bird. The American naturalist had found a female at the head of her brood, in the forest. As he approached, he saw her feathers bristle and heard her hiss menacingly, like a goose. Meanwhile, the ducklings were scampering off in all directions. His highly trained dog brought them all back to him, one by one, without harming them. But his every move was being watched by the mother, who kept walking across his path, as if to hamper his research. When the ducklings were all in the game-pouch, squeaking and struggling, the mother came and sat near the hunter, her face full of sadness. Overcome with pity, as he watched her roll almost under his feet, Audubon handed back her young family and walked away. "As I turned around to look at her," Audubon wrote, "I really believe I saw an expression of gratitude in her eyes; it was one of the happiest moments of my life".

Many examples show that wild ducks are capable of becoming attached to man. They can be tamed quite easily.

Domestic ducks *(Anas Domestica)* are descended from wild ducks. The first domestic duck, the forefather of a family which has grown to a prodigious size, must certainly have come from an egg taken from the reeds in a marsh, and then given to a hen for brooding.

Ducks, which have been domesticated for a very long time, are an important culinary item, and have an honored place in the farmyard. Their eggs are nourishing and pleasant, and their meat is delicious.

All the peoples of the world raise ducks, but the Chinese are unquestionably the masters. They have a way of using artificial heat to make them hatch and have bred some superb varieties.

The Common Duck which we have just described is the typical species of duck; but there are about 70 other species as well. The most noteworthy of these are: the Pochards, the Eider, the Teal, the Scoter and the group known in Latin as Mergus.

The Pochard *(Anas Ferina)* is the most widespread variety in Europe, after the Common Duck. Only slightly smaller than the latter, it nests among pond reeds, and feeds almost entirely on worms, molluscs and small fish. Its nearest American relative is the Redhead.

The Eider *(Anas Mollissima)* is the bird from the north which provides the down celebrated for its softness, lightness and warmth which has given us the word eiderdown. Its plumage is whitish, but the crown, belly and tail are black.

The Eider inhabits the icy seas of Northern Europe, feeding on fish and seaworms. It nests among rocks at the sea's edge. Sometimes two females brood in the same nest, which means that the nest will have at least 9, possibly ten new occupants, as each female lays about 5 or 6 eggs. The nest is crudely built out of marine plants, but is lined inside with a very thick layer of down, which the bird plucks from its own belly.

The Teal *(Anas Crecca)* is smaller than the Common Duck. The Romans had domesticated this bird, as appears from Colu-mellus *(De re rustica)*. But today the teal is wild again, which is a regrettable loss for the farmyard. Teal meat is highly prized.

The Scoter *(Anas Nigra)* is almost as big as the Common Duck, but shorter and more compact. Its plumage is entirely black, though a greyish color in youth. The Scoter spends its entire life on the water, coming ashore only when driven there by a storm or in order to nest in the marshes. It flutters, rather than actually flying across the water, using its wings only to escape danger or to get from one point to another a little more quickly. Its feet, however, constantly drag along the top of the waves, as if it were reluctant to leave its favorite element. Like the petrels, it has the singular privilege of being able to run across the waves. It inhabits both continents.

(A/B/C)

Ducks: Mergus

Birds bearing this Latin name (from *mergere*, to submerge) are sometimes thought of as a family separate from the Ducks; they have a distinctive slender, almost cylindrical, bill, the edges of which are armed with spikes sloping backwards, rather like the teeth of a saw. Yet the carriage, feathers and habits of these birds are quite similar to those of the Ducks.

The birds which make up this family, such as the Smew, the Red-Breasted Merganser and the Goosander, are rarely seen on land; they are exclusively water-birds, living in rivers, lakes and ponds. The Romans used to call them *Mergus*, on account of their habit of swimming with their body underwater and only their head showing. They are voracious fish-eaters. Since they can store a vast amount of air in their trachea, they can last quite some time without breathing. They put this faculty to good use by diving to the bottom, looking for food, and popping up, some time later, many yards away. Their meat has an unpleasant taste.

(D)

(A)

(B)

(C)

(D)

Lapwing

The upper part of the Lapwing's bill is swollen, with the orifice of the nostrils running two-thirds of its length; they have very short toes and pointed wings. In flight, they make a slow flapping noise.

Lapwings migrate from the North in huge swarms early in the fall, flying back in the spring. They live near swamps and ponds, in fact anywhere with the damp soil in which earthworms, insect grubs and slugs abound. They can often be seen descending on freshly plowed land, where they find ample supplies of earthworms. They use a very ingenious method to lure their victims out of the ground. They shake the ground by tapping on it with their foot, thus making the worm think a mole is coming; it wriggles to the surface to escape from its underground predator, straight into the jaws of the lapwing.

Lapwings are remarkably clean birds. Having spent several hours poking around in wet earth, they fly to a nearby lake to wash their bill and feet, repeating the process two or three times a day.

They are gregarious, except during the mating-season, when they go off in pairs. The female lays three or four eggs in a very simple nest in the marshes, on an unconcealed mound. These eggs are said to be quite delicious. (A)

Woodcock

The bill of the Woodcock is very long, straight, slender and soft, with a swollen end; its head is very close to its body, its tarsi are short and its legs are covered with feathers. It lives in the woods. By nature suspicious and nasty, it hides throughout the day, spending its time turning over the dry leaves with its bill, looking for worms and grubs. Woodcocks shrink away from bright daylight, and can see properly only in the evening or early morning. They then emerge from their dark retreat and go off looking for food in the plowed fields, and damp grasslands or in the vicinity of fountains.

They live alone for most of the year and mate in spring, building their nest on the ground with grass and roots near a tree trunk or hollybush. The female lays 4—5 oblong eggs, slightly larger than a pigeon's egg. The young are able to run as soon as they are hatched; the parents stay with them and, if they are threatened, they put the chicks under their neck and, with the aid of their bill transport them quite a long way.

These birds seem to get very attached to places they have known, and tend to come back year after year.

The woodcock's plumage is an extraordinarily harmonious blend of shades: brown, russet, grey, black and white. It is not uncommon to find completely white woodcocks, while others have grey or brown patches on a white background.

The woodcock is a very clean bird which would never dream of going to sleep or getting up without completing its toilet. Every morning and evening, it can be seen flying swiftly towards fountains or streams to quench its thirst and to wash its bill and feet. (B)

(A)

(B)

(A)

Plovers

The main difference between Plovers and Lapwings is that Plovers have no big toe. Otherwise, there are many points of similarity: the same bill, the same habit of assembling in large flocks in damp places, of catching and feeding on worms; both Plovers and Lapwings wash themselves frequently; they associate a great deal, particularly when travelling.

The *Dotterel* is about the size of a blackbird.

The *Killdeer* is about half the size of the Dotterel, and is easily distinguished by its black striped collar and its golden and extraordinarily brilliant eyes.

(A)

Sandpipers

The *Sandpiper*, with its long slender legs, is a graceful and sprightly bird. It is a delight to watch, as it moves swiftly over the wet sand or the swampy grasslands, looking for the worms, small crustaceans and molluscs which make up its diet. Sandpipers are peaceful, gregarious birds. They assemble in large flocks, except at breeding time, and migrate in autumn and spring. The *Common Sandpiper* nests in a wide variety of regions, often passing completely unnoticed as it tends to live in seclusion, near the water's edge. It makes a very pleasant trill sound during the mating season. There are many other species of sandpiper.

(B)

Curlew

The extraordinary length of the bill of the Curlews is their most remarkable feature; the bill is also slender, curved and round for its whole length. Their wings are rather small, and their tail is short. Their plumage is a mixture of grey, russet, brown, fawn and white. Their name is derived from the sad and slow cry which they emit when taking to the air.

These birds inhabit the seashore, near marshes and wet grasslands, feeding on worms, water insects and small molluscs. They dig their bill into the ground, at the same time shaking it enough to disturb the worms who promptly rise to the surface and are eaten up. Normally, curlews walk in a rather measured, solemn way; but, when disturbed, they start running very fast and take off. They are capable of sustained flight, but rarely venture inland; they are always found on the coast. Except during reproduction, they live in large flocks. They are wild and distrustful by nature. Curlews can be found all over the world.

(110)

(C)

(A)

(B)

Bustards

Bustards have a short bill, a stout body, long tarsi and partly bare legs. They have short toes, with the big toe virtually missing, and can run extremely fast, using their wings to help propel themselves along. Their flight, on the other hand, is slow and ungainly. They live in arid, open country, nesting on the ground. They feed on worms, insects, grass, even on seeds, and fly in large flocks over short distances. The males, being more numerous than the females, indulge in polygamy. Bustards are shy and fearful. They are a highly prized game-bird.

The *Great Bustard* is the biggest bird in Europe, weighing as much as 35 lbs. It has a yellow back, with black stripes; its front is a greyish white. The male has curly feathers, like moustaches, on either side of its head. It flies with great difficulty, and can bring itself to take off only when absolutely required to. It lays two or three eggs among grass or wheat stalks, in a nest which is little more than a hole in the ground, very scantily lined inside. (A)

Cranes

Cranes have a moderate gape, a short big toe which does not touch the ground and long, pointed wings. Since they are essentially migratory birds, they combine very powerful flight and the ability to go without food or water for several days.

The Crane (Grus grus) is a magnificent bird, standing nearly 5 feet tall. Apart from its neck, which is black, all parts of its body are a uniform ashen grey. It has a noble, graceful carriage, and its rump feathers, which stand up in undulating tufts, are particularly striking. The Cranes reach Europe in April or May and spend the summer in the far north. At the onset of the cold weather, in mid-October, they leave to spend the winter in Egypt, Ethiopia and even in South Asia. They travel by night in flocks of variable size, sometimes comprising 200—300 birds, all flying in a long pointed V-shape, with the point foremost.

Cranes deserve their reputation as a symbol of vigilance; when they are asleep, with their head curled up under a wing, one of them stays on watch and is able to alert the whole flock to danger.

If captured young, they are easily tamed, and become very familiar quite quickly.

The Demoiselle Crane is easily distinguished by a large crest-like tuft of feathers behind each ear, and by a hanging black tuft which adorns its breast. It is the same size as the crane described above, with even more elegant shapes. It also has more of a gift for mimicry. Its slightest gesture is poised and sophisticated, as if it was trying to draw attention to itself. It may be found in Turkey and Southern Russia, in North Africa and in certain parts of Asia. The Crowned Crane has a splendid sheaf of feathers adorning the top of its head; it can open them up like a fan, to make a resplendent diadem. It is a slender, graceful bird, quite as tall as its two sister-species, but with a most striking voice. It seeks out the company of man, and quickly becomes quite familiar. It can be found on the eastern and northern shores of Africa. (B)

Sunbittern

The Sunbittern is about the same size as a partridge, with a wide, outspread tail. On account of its beautiful colors, it is known, in Guyana, as the Little Peacock of the Streets, and Sunbird. It is very wild. (C)

(B)

(A)

(A)

(C)

Moorhens (U.S.: Florida Gallinule)

The moorhen's most striking feature is its short, stout bill, which begins with a frontal shield on the bird's forehead, is thick at the base and ends in a point; also its four toes, which are highly developed and equipped with sharp talons. It is found in marshy country and on lakes and rivers, where it feeds on worms, insects, molluscs and small fish. Moorhens are sprightly, graceful birds, who spend the day among the reeds or striding across the broad leaves of the water lilies, emerging from their retreat only in the morning and at night, to look for food.

Though residents in some countries, they are migrants in others, travelling by land, sea or air. They take the same route year after year, and nest consistently at the site of their first brooding.

The male and the female take it in turns to incubate their seven or eight eggs and never fail to cover them over with grass whenever they have to leave the nest.

The *Purple Gallinule* is a sort of exaggerated version of the moorhen. Its bill is thicker and stronger, and its frontal shield larger; it also has longer toes. Yet its habits are the same. However, as it has a weakness for cereals, it tends to spend less of its time in the water. It is a superb creature, with entirely indigo blue plumage and a pink bill and feet. (A/B)

Sandgrouse

Sandgrouse have a short bill, a small head, a bulky and round body, a short tail, bare tarsus with spur varying in size, and a medium-sized big toe. They are built for long flights, having long and pointed wings; yet, in fact, they never travel very far. Like pigeons, they have a high, fast and sustained flight. They reside in the arid plains of Southern Europe, Asia and Africa. (C)

(A)

(C)

(B)

Turtle Doves and Barbary Doves

The *Turtle Dove* occurs throughout Europe, but is commonest in the South. It reaches France in the spring, and leaves for warmer climates at the end of the summer. When nesting, it prefers darkness and seclusion, which it finds in the bigger trees. It feeds on seeds and berries. After harvest-time, it frequents fields where wheat or other cereals have been growing, putting on much weight in the process and thereby becoming a succulent game-bird. Although it is a very wild bird in the natural state, the Turtle Dove can be tamed quite easily if caught young enough, and then can be very friendly.

The *Barbary Dove (Streptopelia risoria)* came originally from Africa, where it still lives in the wild. It is this species which is raised in cages and aviaries. In some towns in Egypt, these doves are so tame that they walk in the streets, and even go inside people's houses, without any fear of man. Their fertility is great, as they breed once a month, except during their molt. Their voice is a monotonous, plaintive cooing sound, which has some of the quality of laughter, whence the Latin name *risoria*.

For the ancients, the dove was a symbol of tenderness; one cannot deny that the male and the female are extremely affectionate to one another. (A)

Pigeons and Doves

The *Wood Pigeon* is the largest species of the *Columbidae*; it has a fine, slender bill, long wings and short tarsi. It has a slate-grey plumage, with shades of blue, green and pink. It occurs throughout the warm and temperate zones.

Wood Pigeons live in forests and are particularly fond of the tops of very tall trees. Strawberries are a great delicacy for them, though their diet consists mainly of acorns and beechnuts. When these items are in short supply, they descend on farmlands and try to unearth the freshly sprouting cereal crops with their bill, thereby causing much damage. Having chosen the site for the nest, in a tree with dense foliage, the female then sets about the work of building it, using materials brought to her by the male. These are always twigs which the pigeon snaps off using his bill or feet; he never collects dead twigs lying around on the ground.

In their natural state, wood pigeons are wild and hostile, but their character changes as they come into contact with man, (B)

(B)

(A)

(A)

(A)

either in the domestic state or near human settlements. When taken very young, these birds become friendly with no trouble at all, and do not seem to miss their lost freedom.

The *Stock Dove* looks very like the wood pigeon, but is smaller; its habits are the same, but for the fact that it builds its nest in hollow trees, instead of on branches.

The *Rock Dove* prefers an arid, rocky habitat, but seems quite willing to sacrifice its independence to live in the special little houses, called *dovecotes*, which man has made for it. It is commonly thought to be the origin of many breeds of domestic pigeons.

The *Carrier Pigeon* is a small bird, with great facility and great speed in flight. It is noted for its devotion to its place of birth and the place where its offspring were raised, and also for the remarkable intelligence which guides it back home. These birds have been taken far from home in closed baskets and then set free; they were found to return to their point of departure, after varying periods of time, without any hesitation. This valuable trait has long been exploited, especially in the Orient.

Crested Curassow

In both shape and size, the Crested Curassow resembles the turkey, and, indeed, constitutes the South American branch of the turkey family. They have no spurs, and wear on their head a large crest made of erectile and curved feathers. Large flocks of them live in the forests, looking for the seeds, berries and buds which make up their diet. As they are gentle by temperament, they readily become domesticated and quite familiar. Curassow meat is delicious. (A)

Turkey

The Turkeys are big birds. Their head and neck are bare, but decorated with fleshy appendages; the one on the head, which hangs forward over the bill, may swell and become erect under the influence of anger or sexual arousal. A cluster of long, stiff hairs, like horsehair, hangs at the base of the neck. Its tarsi are robust, and have a rather undeveloped spur; the tail is rounded, of medium length, and can spread into a fan.

The Common Turkey is originally from North America, where it still lives in the wild. It is frequently found in the forests along the shores of the Mississippi, Missouri and Ohio Rivers.

The color of the wild turkey is brown, with some irridescent blues and greens, and a metallic sheen. The male is usually 4 ft. 3 inches tall with an average weight of 17—19 lbs. Audubon claimed to have seen one weighing 40 lbs. The female is much smaller, weighing barely more than 11 lbs; her plumage is less colorful than that of the male. Naturally, such a heavy bird flies only with some difficulty. The wild turkey can fly quite a long way, but usually flies only when all other means of locomotion are out of the question. However, it does run with astonishing speed. In fact, it can outrun the fastest dog and allows itself to be caught only after a chase lasting several hours. It travels long distances on foot; these journeys are not of a periodic nature, and seem to be determined by a lack of food in the area it normally inhabits. (B)

Guineafowl

Guineafowl may be distinguished as follows: a small head, short neck and bill, short and hanging tail, very short tarsi without spurs, a round body, and short, concave wings. They have a callous comb on top of their head, of a reddish blue color, though this is sometimes replaced by a tuft. Fleshy wattles hang under their bill. The Common Guineafowl has a slate-grey plumage, with white patches. This bird, originally from Africa, was well known to the Greeks and Romans. Indeed, in Greece it became a symbol of fraternal devotion. Yet it is a rowdy, quarrelsome bird, very quick to attack the young of other birds, and split open their skull with a single jab of its bill. (C)

Partridges

Partridges have a very curved bill, a stout body, blunt wings and a short, trailing tail. They live on the ground, and perch only when absolutely required to. They can run amazingly fast and effortlessly. Their flight is also quite fast, but they cannot fly very far or more than a short distance above the ground.

Partridges are residents, rather than migrants, and live in groups. They are monogamous, keeping the same partner from one spring to the next.

In certain species, such as the *Red-Legged Partridge*, in which the females outnumber the males, a number of males necessarily remain unattached. This leads to fights between husbands and bachelors, as the latter, unhappy with their lonely state, try to oust their neighbor and have a family of their own. Sooner or later, however, the fighting ends, and the unlucky ones withdraw to form all-male groups.

Partridges are by nature timid and fearful, as their behavior frequently shows; they seem to see enemies everywhere. Such distrustfulness is hardly surprising when one considers that, not counting man, the fox and the birds of prey wage a full-scale war against them. They find the birds of prey particularly terrifying; as soon as they spot one, they stand still, petrified, and roll up into a ball, completely motionless, hoping not to be noticed. (D)

(B)

(C)

(D)

(D)

(A)

(D)

Francolin

What distinguishes the Francolins from the partridges is their tougher and longer bill, their more highly developed tail and the one or two very sharp spurs to be seen on the males. Their habits are also different. They live in wooded, swampy areas, feeding on berries, seeds, worms, insects and young bulbed plants. Another peculiarity is their habit of continuously perching on trees and never spending the night on the ground. Apart from these distinctive traits, they live exactly the same way as partridges. (A)

Ptarmigans

The feet of the *Ptarmigan* and the hare are rather similar; whence their scientific name *lagopus*, which means *hare-foot*. In addition to having feathered tarsi, these birds also have fur on their toes, right down to the base of their claws, just like a hare.

In both Old and New Worlds, they are to be found in the icy zones and around the peaks of the high mountains. They love the snow, and delight in rolling around in it. They dig holes in the snow with their feet, and climb in to spend the night or protect themselves from gale-force winds. Except for a black line on their face and a number of dark flight feathers, their plumage is a brilliant white. White is their winter coat. But in summer, when the hot sun has melted all the snow, they put on a greyish plumage, mottled with brown and rust patches. (B)

Quail

Quails have a small bill, a short big toe, a rudimentary spur, in the form of a horny knob on their tarsi, a stout body, medium-size and pointed wings, with hardly any tail at all. There are several species of quail.

The Common Quail is famous for its migrations. Every year, vast flocks of them leave the remotest parts of Africa, cross the Mediterranean and, early in May they appear all over Europe. They cover this enormous distance in the opposite direction in September.

Quails travel in the evening and at night. They fly at a fairly high altitude, but never against the wind. They try, on the contrary, to have the winds blow them across the Mediterranean. In this way, they come in on the southerly winds and go back to Africa on the winds from the north. If they encounter a storm on their way, they are not strong enough to go on through it and fall by their thousands into the sea.

Quails inhabit harvest-covered plains and fertile grazing lands. They love to frolic in the dust and never perch. They feed on seeds and insects. They are not very sociable: the sexes come together only in the mating season and separate as soon as the young no longer need maternal care—which is very soon, because the young develop very fast. The females lay eggs twice a year, once in Africa and once in Europe, laying 10 to 14 eggs each time. (C)

Monal Pheasant

These birds are very fond of cold climates; for instance, they seem particularly to favor the highest peaks in the Himalayas. One species in India is so resplendent in its coat of brilliant colors that it is known as the Bird of Gold. (D)

Grouse

Grouse inhabit the pine and birch forests of high mountain regions. Certain species prefer heather-covered moors. They feed on a great variety of things: fruit, berries, pine and birch buds, insects and small earthworms. These birds have a proud, warlike bearing, with a sturdy figure, black plumage with white patches and bluish reflections. They are polygamous and live in families. They readily take refuge in trees, either to sleep or to escape from their enemies.

From there, as soon as spring begins, the males' extremely discordant voice can be heard calling the females. Every morning and every evening, at dawn and dusk, this frightful cacophony can be heard up to several miles away. (E)

(C)

(E)

(E)

(D)

(A)

(B)

Pheasants

The Pheasants, though originally from Asia, are now to be found all over the world. Their most striking feature is their extremely long tail, the middle quills of which may be as much as 5 feet long. They are also noted for their dashing figure, their elegant shape and their brilliant plumage—brilliant, that is, in the case of the males, since the females are much more modestly adorned. Their cheeks and the area around their eyes are bare and covered with a hard, wartlike substance. The males wear a spur.

Pheasants live in flat country, in woods or swamps. Their varied diet consists of seeds, berries, worms, insects and snails. Being very wild by temperament, they fly away at the slightest sign of danger. They live alone until brooding time, when the males, being polygamous, set about composing their harem. Savage fights take place for possession of the females, the weaker birds sometimes getting killed in the process.

Pheasants lay their eggs on the ground, in the midst of thick undergrowth, between 12 and 24 at a time. The remarkable maternal devotion so characteristic of most birds is curiously lacking in this species; the mother hardly seems to know her own offspring, and simply cares for whatever young pheasants happen to be near her.

A very peculiar feature of this family of birds is that when the females reach the age of infertility, they acquire the voice and the plumage of the males.

The Golden Pheasant and the Silver Pheasant are two splendid birds, originally from China and Japan. The first, with a plumage of purple and gold, has a beautiful golden crest on its head; the second, whose coloring is black and white, is certainly its equal in beauty.

(A)

Peacock

The main characteristic of the *Peacocks* is the immense, splendid tail bestowed on them by Nature. It is made up of long, wide, thick feathers, very richly colored, and, like the turkey's tail, is erectile. The Peacock has a mantle on which purple and gold blend with the most delicate tones of irridescent emerald; its plumage is covered with countless shining eyes; it has a tall, dignified posture, an elegant body, a noble carriage, and its head is crowned with a slender, floating crest, the symbol of royalty; it is no wonder, then, that men have always thought of it as the most beautiful of all the birds.

The Peacock was known early in ancient times; the Bible mentions it as one of the most precious goods brought back from Asia by King Solomon's fleets. The bird made its appearance in Greece as a result of Alexander's Indian expedition. He is said to have been so over-awed by the sight of this creature that he imposed the most severe punishments on anyone who killed it.

The Peacock can run so fast that it is frequently able to escape from dogs pursuing it; its take-off and flight, however, are far from effortless, though it is capable of flying quite a long way. Its diet consists of a wide variety of seeds, which it swallows whole. In the evening, it perches at the top of the tallest trees.

Even when domesticated, this liking for high places persists. Peacocks simply love the roofs of houses, which they often damage quite badly, either by upsetting the tiles or by eating the thatch, as the case may be. They have a strong destructive instinct.

From time to time, the peacock lets out a deafening cry, which contrasts rather disagreeably with its beautiful plumage.

The peacock is polygamous. Early in spring, the male tries to dazzle the female with his splendid display of feathers; he struts to and fro, with all his tail-feathers fully outspread, taking an obvious delight in the impression he is creating and contentedly noting the cries of admiration which can be heard from his admiring female audience. His vanity is truly limitless. Late in August, he loses his beautiful feathers, and regains them only the following spring.

Like the females of roosters and pheasants, the pea-hen acquires the plumage of the male when she reaches the age of infertility or when a premature atrophy of the ovaries has made her sterile. The lifespan of the peacock is between twenty and thirty years; those writers who have claimed that it is as much as a century are wrong.

(B)

(A)

(B)

(B)

Ostrich

The head of the Ostrich is bald and callous, with a short bill, pointing downwards and rounded at the tip; its legs are half-bare, very muscular and fleshy; its tarsi are long and thick, ending in two toes facing forwards, one of them, the shorter of the two, having no claw; the very short wings are made up of soft, flexible feathers; the tail is in the form of a tuft.

This order comprises only one species, which is very widespread in the hinterland of Africa, down as far as the Cape of Good Hope. This bird usually stands 6 feet tall, but may reach a height of 10 feet 8 inches; it weighs between 88 and 110 lbs.

The ostrich was known in the earliest times. It was mentioned in the Bible; Moses forbade the Hebrews to eat ostrich meat, considering it unclean, whereas the Romans, on the other hand, regarded it as a great delicacy. In Imperial Rome it was always abundant; the extravagant Heliogabalus went so far as to serve 600 ostrich brains at one banquet.

The ostrich is extremely greedy. While its sight and hearing are highly developed—it can hear the faintest sound and is reported to be able to see up to 5 miles—its taste and smell are quite imperfect. This explains why it tends to pick up anything it sees. In the wild state, it swallows quite large stones in order to enhance the digestive powers of its stomach; in captivity it will swallow bits of wood, metal, glass, plaster, chalk, etc. It is nevertheless capable of going without food and water for several days.

It has astonishing muscular strength. When domesticated, it carries a man on its back with ease; it can quickly be trained to haul loads or allow itself to be ridden, like a horse. The tyrant Firmius, who reigned in Egypt, in the third century, used to have himself drawn by a team of ostriches.

Upon being mounted by a rider, the ostrich begins to trot; soon it gets warmed up and, stretching its wings, starts to run so fast that it seems to be flying across the ground. Its kick alone is sufficiently strong to protect it against any animal roaming the desert. Its foot is such a powerful weapon that one well-aimed kick in the chest can kill a man.

Ostriches are eminently sociable animals: sometimes they can be seen in the desert in flocks of two or three hundred, mingling with herds of zebras, etc. They mate towards the end of autumn.

The nest of an ostrich is more than 39 inches in diameter. It is just a hole in the sand surrounded by a wall made of soil taken from the earth; outside it, is a ditch for drainage of water.

Each female lays between 15 and 30 eggs, depending on the circumstances. The weight of the eggs varies from 2.2 lbs to 3.3 lbs, each one being the equivalent of 25 hen's eggs. They have an excellent taste; one single egg could provide an ample lunch for two persons.

In spite of their great strength, ostriches are very peaceful and are easily domesticated. (A)

Cassowaries

The Cassowaries are related to the ostrich, but differ on account of some very distinctive features. Their shape is less elegant, and they are even less well equipped than the ostrich for flight. Their much shorter wings are no help even in running. Their long feathers, blackish in color and almost entirely lacking in barbs, look rather like a horse's mane; they have three toes on each foot.

The Helmeted Cassowary carries on top of its head a sort of helmet which is really a swelling of the skull bones, covered with a horny substance. This massive bird inhabits the islands of the Indian Archipelago, the Moluccas, Java, Sumatra. It is particularly abundant in the dense jungles of the island of Ceylon. It feeds on grasses, fruit and sometimes on small animals. It is fierce, strong and brutal; one arouses its anger at one's own risk. Though very short, its wings are a good defensive weapon, as each of them is equipped with five sharp spikes, the middle one of which is a foot long. Its normal cry is a low grunting sound which becomes a snorting, buzzing noise, rather like the noise of a car or of distant thunder, when it becomes angry.

The New Guinea Cassowary stands taller than the one we have just described, also having no helmet, wattles or spikes on the wings. (B)

(B)

(A)